The Geology of

ACADIA NATIONAL PARK

Opposite: East side of Mount Desert Island looking north. To the left is Dorr Mountain and the eastern flank of Cadillac Mountain. At right center is Great Head and Sand Beach. Near the top are the Porcupine Islands and Frenchman Bay.
Reverse: Geologic map of Mount Desert Island.

The Geology of
ACADIA NATIONAL PARK

Carleton A. Chapman

Published by **THE CHATHAM PRESS, INC.**
Distributed by **THE VIKING PRESS, INC.**

Photographic credits: Cover photo by Maxwell Rollins courtesy Down East Magazine. Air views of Mount Desert Island and Isle au Haut are by Laurence R. Lowry. All other photos are by Mr. Frederick J. Pratson or through courtesy of the National Park Service.

Produced in cooperation with the United States Department of the Interior, National Park Service, Acadia National Park, Hulls Cove, Maine.

Library of Congress Catalog Card No: 73-107079.

SBN 87638-012-7

Manufactured in the United States of America by the Eastern Press, Inc., New Haven, Connecticut.

ACKNOWLEDGEMENTS

The field work and laboratory study involved in the geologic investigation of the Maine coastal region were financed by grants from the Research Board, University of Illinois, and the Eastern National Park and Monument Association. Dewey Amos, Carl Chapman, William Chapman, Forest Etheridge, Robert Luce, Jackie McGregor, Terry Offield, Michael Schneider, Rubini Soeria-Atmadja, Jack Wehrenberg, Eugene Williams, and Paul Wingard assisted in the field study. Paul Favour, Jr. (former Chief Park Naturalist, Acadia National Park) and Richard Smith spent several days in the field with the author and assisted materially in the preparation of the manuscript. Paul Favour, Jr., Bennett Gale, John Good, Robert Rose, and others of the National Park Service have offered many valuable suggestions and have read the manuscript critically. To these individuals and organizations the author is deeply indebted.

33665

CONTENTS

Otter Cliffs.

FOREWORD

This book presents in relatively nontechnical language the fascinating story of the geologic creation of the Maine coastal region, home of Acadia National Park. It explains the natural forces and processes that produced the numerous enchanting and captivating features of the landscape. A better understanding of these phenomena can lead to an increased appreciation of the area from both a scientific and aesthetic point of view. There is, perhaps, no place in the world of comparable size which is more accessible and where a wider variety of geologic features may be more clearly observed than in Acadia National Park. Although a general knowledge of these features may be gained from reading the first section of this book, a deeper understanding will be obtained by those who take time to visit and observe the areas described in the succeeding sections.

In so brief an account as this it is possible to present only a simplified and generalized picture of the local geology. For most Park visitors, however, such a treatment is adequate. Those interested in more detailed and technical accounts will find a list of supplementary readings in Appendix II.

The numerous and detailed self-guided trips will serve as a valuable field guide, not only for the layman but also for secondary school or more advanced students of earth science. Special pains have been taken in organizing these rides and walks, including consideration of such matters as conservation of time, personal safety, and respect for private property. The excursions cover the best examples of the geologic features and phenomena, and have been arranged to provide the most scenic views, suitable parking sites, and practical traffic patterns. By traveling as indicated in each self-guided route, one may readily follow the detailed descriptive material.

PART I

THE GEOLOGIC STORY OF MOUNT DESERT ISLAND

1. FORMATION OF THE STRATIFIED ROCKS

Ellsworth Schist.

Our story begins about 450 million years ago when the sea covered a large part of New England and widespread layers of sand, silt, and mud were accumulating upon its floor. Most of this material was being carried as sediment by the streams that drained adjacent highlands. The tiny particles were derived in part from the solid rock ledges by weathering and disintegration. Much material, however, is believed to have been blown into the air as small particles during explosive volcanic eruptions. Some of this settled directly to form irregular layers of volcanic ash and tuff, but much was reworked by waves to form well-bedded or stratified deposits.

After these thinly laminated and well-bedded rocks had accumulated to a total thickness of hundreds and perhaps even thousands of feet, the southeastern portion of Maine became an unstable region of the earth and crustal movement (diastrophism) set in. During this disturbance the stratified (layered) rocks were compressed, heated, and transformed to metamorphic rocks. The transformation or metamorphism involved extensive change of rock material as the tiny particles were made over or recrystallized to form coarser grains. Water trapped between grains of the original sediments probably aided greatly in this recrystallization process by promoting the solution of certain rock constituents and the precipitation of others. The numerous closely-spaced planes of bedding provided channelways along which solutions could readily move and concentrate certain minerals into thin layers and films, thus giving to the rock its characteristic foliation (layerlike quality) and tendency to split into thin leaves and slabs.

These metamorphic rocks in the park area constitute the rock formation known as the Ellsworth Schist and consist principally of two rock types; schists that are rich in mica and split readily along the surfaces of foliation and gneisses which resist splitting because of their more poorly developed foliation.

After folding and metamorphism, the Ellsworth Schist was elevated to form dry land, but through extensive erosion the area was eventually reduced to near sea level. The folded rock layers were locally truncated by the new land surface so that the amount and direction of their inclination appear to vary widely from place to place (Figure 1A).

Surf at Thunder Hole.

Cranberry Island Series.

The next rock unit, known as the Cranberry Island Series, is believed to be about 420 million years old. It was laid down upon the crumpled, metamorphosed, and eroded layers of the Ellsworth Schist. The actual contact between the two units has not yet been found on Mount Desert Island, but it is believed to have been observed about 15 miles to the west (Figure 1B).

This series is composed predominantly of volcanic rocks with smaller quantities of inter-bedded sedimentary rocks. The volcanic rocks consist largely of fine ash and rock fragments which were formed during repeated explosive eruptions and settled upon land and sea. Most of the fragments are less than an inch across, but blocks up to many feet long have been seen. These fragmental materials have been hardened to a compact durable tuff (volcanic rock). Since much of the series is well bedded or stratified, it is believed that the volcanic materials were extensively washed and sorted by the waves. Some lava flows (mostly felsite) were interleaved with the fragmental volcanic deposits.

Perhaps several thousand feet of material accumulated before crustal unrest and erosion again set in. During this deformation the volcanic layers were locally folded so that now some are seen posed in a nearly vertical position. During this disturbance metamorphism must have been mild because recrystallization of the rocks was not pronounced (Figure 1C).

Bar Harbor Series.

Following this second period of crustal disturbance and erosion, the land surface was again reduced nearly to sea level. Depression occurred in the vicinity of Mount Desert Island, and a shallow sea spread over the land. Large fragments of older rock, dislodged from the nearby ledges, were tumbled and rolled about by stream and ocean water until their corners were completely worn away and rounded pebbles and cobbles formed. These rounded fragments were locally concentrated and more or less mixed with coarse sands to form gravel deposits along old shore lines and stream channels. Thus, the soil and partly disintegrated material that mantled the newly flooded land surface were transformed by stream and wave action to a blanket of gravelly material on the new sea floor.

On top of the gravel and coarse sand were deposited layers of finer sand and silt totalling many hundreds of feet in thickness. This entire sequence of beds is known as the Bar Harbor Series. The rocks are predominantly brown, gray, or green siltstone and split along bedding

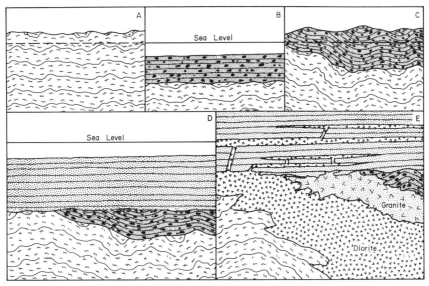

Figure 1. The sequence of rock formation. A.) The Ellsworth Schist was deposited as horizontal layers of sedimentary and volcanic material, then crumpled and metamorphosed, and finally uplifted and eroded. B.) The Ellsworth schist was submerged and the Cranberry Island Series then was deposited. C.) Folding, uplift, and erosion occurred again. D.) Submergence was followed by deposition of the Bar Harbor Series. E.) Dioritic magma invaded from below, spread out, then solidified as great sheets of diorite in the stratified rocks. Granitic magma then invaded and crystallized to form the fine-grained granite.

planes like flagstone; the gravelly material, now hardened, is known as conglomerate (Figure 1D). Some of the whiter layers are composed of volcanic ash, much of which may have fallen directly into the sea and been gradually reworked by the waves.

Thus far we have recounted the formation of three distinct units of superimposed stratified rocks. Each younger and overlying unit is separated from the next older unit below by a distinct break. Sedimentary and volcanic rocks of the oldest unit (Ellsworth Schist) were metamorphosed to schist and gneiss, uplifted and eroded, and buried beneath thousands of feet of volcanic rocks of the next younger unit (Cranberry Island Series). During a second period of crustal disturbance, both rock units were somewhat folded, mildly metamorphosed, uplifted, and eroded. A third time the land surface was depressed beneath the sea; and a thick sequence of well-bedded silty rocks, known as the Bar Harbor Series, accumulated.

This is the first phase of our story, that of sedimentation or the accumulation and formation of the three series of stratified rocks of the Mount Desert Island region. Such was the framework of the upper part of the earth's crust in Devonian time (about 375 million years ago) when the second phase of our story, that of magmatic invasion, began. This phase has to do with the origin of the diorite, various types of granite, and related crystalline rocks.

2. GREAT INVASIONS OF MOLTEN ROCK

Thick Sheets of Diorite.

Very little change in the crustal level is believed to have occurred during the interval between the deposition of the Bar Harbor Series and the first major invasion of molten rock material (magma) from great depths. The hot fluid, similar to the black lava of Hawaii's volcanoes, was squeezed upward and sought the easiest means of access to the stratified rocks above. Zones of weakness, such as large fractures in the crustal rocks, permitted rapid influx of the magma. It broke across the more deeply buried layers of rock and advanced rapidly to higher levels in the Bar Harbor Series where it was able to spread out horizontally in the form of flat sheets by wedging its way along the bedding planes of the rock layers (Figure 1E). It seems likely that some material found its way to the earth's surface and poured out as lava from volcanoes. The layerlike masses found in the Bar Harbor Series are known as sills, and some are 100 feet thick.

The largest such body, roughly 3,000 feet thick, appears to have formed mainly below the Bar Harbor Series and above the Ellsworth Schist; but the exact relations are not too clear. Very likely it is localized along an old break cutting across the Cranberry Island Series. It is somewhat irregular when compared with the sills within the Bar Harbor Series and will be referred to as an intrusive sheet. (Figure 1E).

The hot magma coming into contact with cold stratified rocks brought about two striking effects. First, the colder rocks were heated and intensely baked so as to form a recrystallized, dense, brittle material known as hornfels. Only a thin zone of hornfels is found against the smaller bodies of injected materials, but adjacent to the larger bodies hornfels zones are many feet thick. One further indication that the magma must have been extremely hot is the generation within the hornfels of minerals known to develop only under high temperatures. Similar heating effects were produced upon the numerous slabs and fragments of stratified rocks that were accidentally incorporated in the rock melt.

Jointing and sheeting in granite along Cadillac Summit Road. Sheeting fractures are nearly parallel to original ground surface. Many vertical jointing surfaces are weather-stained.

The loss of much heat to the cold surrounding rocks caused the magma to congeal by crystallizing. The smaller or thinner masses of magma cooled rapidly and were forced to crystallize with a very fine-grained texture. The flow of heat from the larger bodies extended over a much longer period of time and permitted the growth of larger crystals that formed a coarser texture. Even in the larger bodies, finer grained textures are found at the contacts with the stratified rocks, denoting a more rapid chilling of the magma. A short distance in from the contacts the crystallization period was somewhat longer and the coarser texture formed.

Congealing of the magma resulted in the development of two principal types of minerals — heavy, dark-colored minerals (olivine, pyroxene, and hornblende) and less dense, light-colored minerals (mainly plagioclase feldspar). The mineralogical composition of the rocks within these thick sheets and sills is highly variable even over distances of a few feet, and actually many different rock types are represented. It seems appropriate here, however, to refer to this heterogeneous rock material as diorite.

Cooling must have been sufficiently slow in the large diorite sheet to permit convection currents to be set up within the magma and to allow some settling of heavy crystals through the liquid magma as they formed. In combination, these two types of movement enabled the early-formed crystals to accumulate on the floor of this intrusive sheet. As the floor was built up by the addition of new crystals, which were more or less sorted by the sweeping convection currents, a distinct banded and layered structure was produced in the diorite body.

Fine-Grained Granite.

Shortly after the dioritic magma had solidified, there was a second major invasion of molten rock material, but this time the magma had the composition of granite. The hot liquid broke through some of the deeper rocks and came into contact with the flat, well-bedded rocks of the Bar Harbor Series (Figure 1E). Contact was also made with the diorite. Accompanying this intrusion was intense shattering of the older adjacent rocks, and great quantities of angular fragments of these broken host rocks were caught up and enclosed in the granitic magma. A reaction between magma and rock fragment took place, and both liquid and solid suffered change. Fragments were partly recrystallized and partly dissolved as they soaked in the hot melt. Their angular outlines became more rounded, and the included masses blended with the somewhat contaminated granite. Near the contact with cold host rock, the granitic melt was chilled and forced to crystallize before much re-action with solid fragments could take place. Here, therefore, angular inclusions with distinct boundaries are the rule. Cracks in the solid host rock were immediately injected with magma, which then solidified for form dikes (narrow veins) ans stringers (long fingers) of granite penetrating deeply into the older rocks.

The fine-grained texture, which appears to characterize this body of granite, indicates that the rock formed relatively close to the earth's surface, where the heat and volatile materials escaped quickly and crystallization was rapid.

Another pronounced gap appears in our geologic record after the formation of the fine-grained granite. By piecing together scattered bits of information from other parts of the state, it seems evident that up-lift and erosion of the land followed the formation of the fine-grained granite. Later, depression again brought the area below sea level and deposition of sediments followed. In late Devonian time (about 350 million years ago) the Mount Desert Island area was probably buried thousands of feet, but subsequent erosion has removed all traces of these sedimentary beds.

Coarse-Grained Granite.

Sometime after the formation of the fine-grained granite, the Mount Desert Island region was being prepared for a complete remodeling of its geologic architecture. In a huge reservoir miles beneath the level of what is now the land surface, granitic magma had accumulated and was working its way slowly toward the surface. The method by which such large bodies of molten rock develop is still one of the great mysteries of geology. The extent of this gigantic pocket of hot liquid was suf-ficient to endanger the solidity of the crustal rocks above it. The weight of these relatively heavy roof rocks upon the light magma beneath was a burden too great to be sustained, and the solid roof of the reservoir began to fracture and sag (Figure 2A). In time it ruptured to form a huge plug-shaped block that sank into magma below (Figure 2B). The displaced magma was forced up to fill the space vacated by the block of fallen roof rocks where it crystallized slowly to form the coarse-grained, pink granite of the area (Figure 2C).

To visualize more clearly the mechanical failure of the roof, we might imagine ourselves to have been looking down from some position high above the magma reservoir. From here we might have seen how

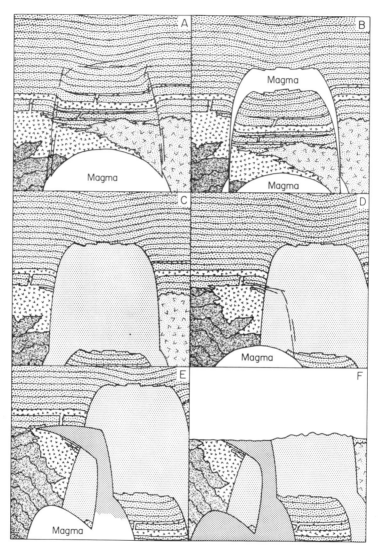

Figure 2. Schematic cross sections showing intrusion of the granites. A.) Sagging and fracturing of the rocks above a magma chamber (clear pattern). B.) Subsidence of a pluglike block into the magma chamber below. C.) Magma fills cavity above block and crystallizes to coarse-grained granite. D.) Fracturing is renewed in rocks above new magma chamber, this time without completely isolating a pluglike block. E.) Flexing and subsidence of incompletely detached block permits upward surge of magma to fill curved cavity. Magma crystallizes to form a ring-dike of medium-grained granite. F.) Erosion strips away overlying rocks and much of the two granites as well, to form the Mount Desert Range.

the sagging of the roof tended to develop great vertical cracks in the subsiding rock (Figure 3A). These vertical fractures were not straight, but from our position above would have appeared as curved or arc-like cracks which outlined the roughly elliptical shape of the underlying magma chamber. As the cracks enlarged and the arcs extended themselves, sinking continued. Smaller fractures gradually joined to form larger ones (Figure 3B); and eventually a plug-shaped block of the crust, overlying the magma chamber but still deep below the land surface, was freed to plunge vertically downward into the hot molten mass below (Figure 3C).

The fallen block did not part from the adjacent rock with a smooth clean-cut break. Instead, a pronounced zone of shattering developed, resulting in a jumbled mass of various sized rock fragments. Such chaotic material is called breccia. Some fragments sank into the melt; others drifted out into the magma and were frozen in as it solidified. The distribution of breccia is shown on the fold-out geologic map by the pattern indicating shatter-zone.

Out beyond the walls of the newly created magma pocket, nearly isolated fractures opened into arc-like gashes or fissures. These were quickly, filled with the hot fluid that crystallized to form long curved dikes of granite (Figure 3C). Such arc-like bodies are known as ring-dikes because, if extended, they would develop complete rings.

The granite of the ring-dikes, as well as that near the contacts of the main body, is noticeably finer textured than that well within the larger mass. This diminution of grain indicates a chilling and rapid crystallization of the magma against the colder adjacent rock.

A baking effect by the granite melt is most pronounced in rocks of the Bar Harbor Series. Near the granite contact these originally fine, silty sediments have been recrystallized (metamorphosed) to a rock called hornfels. This heating effect is indicated by a gradual increase in grain size of the hornfels and by the development of successively new minerals within it as one approaches the granite body. Some appreciation of the intensity and magnitude of this baking action may be gained when we consider the amount of heat liberated by such a gigantic, hot liquid mass. The thermal energy lost to the surrounding rocks by solidification is roughly calculated as equivalent of 170 million tons of fissionable material.

After the magma had crystallized, further disturbance formed arc-like fractures in the main mass of crystallized granite (Figure 3D). Apparently these fractures did not open sufficiently to admit fresh magma but rendered the granite permeable to highly active, hot, watery fluids escaping from greater depth. These restless, potent fluids helped to recrystallize the granite in the vicinity of the fractures. In this change

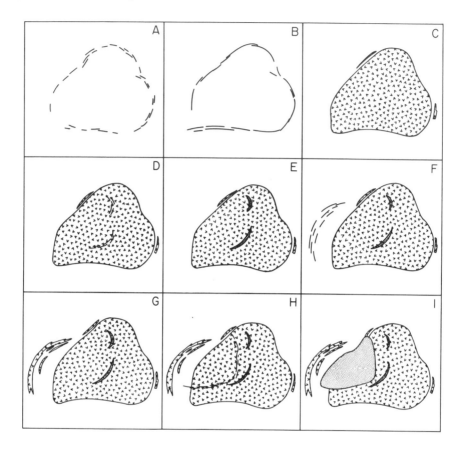

Figure 3. Schematic bird's-eye views showing intrusion of the granites. A.) Curved fracture pattern forms due to subsidence of rocks above large magma chamber. B.) Fractures are extended and subsidence intensified. C.) Pluglike mass collapses into molten rock below. Magma wells up to fill cavity and crystallizes to coarse-grained granite. Fractures beyond the collapsed area open and fill with magma, which crystallizes to form small ring-dikes. D.) Renewed subsidence forms curved zones of intense fracturing within the granite. E.) Hot solutions from below permeate these zones and recrystallize the granite locally to a fine textured rock. F.) Subsidence forms curved fractures west of the coarse-grained granite. G.) Granitic magma fills and crystallizes in the fractures to form ring-dikes. H.) Further subsidence and fracturing occurs within the coarse-grained granite. A second pluglike block becomes only partially detached and flexes downward along its southern and eastern margins. I.) Granitic magma flows up and over sagging block and crystallizes to form ring-dike of medium-grained granite.

most of the large quartz and feldspar crystals were transformed to aggregates of tiny grains, but many remained more or less intact. With its numerous large crystals scattered through a very fine-grained matrix, the changed rock resembles a granite porphyry but may be distinguished from the latter by microscopic examination. In outline these arc-like patches of recrystallized granite have the appearance of small ring-dikes (Figure 3E).

Contemporaneously, perhaps, with fracturing and recrystallization of the coarse-grained granite, arc-shaped fissures formed in the older rocks west of the granite (Figure 3F). Subsequently granitic melt entered this zone of shattering to form numerous narrow ring-dikes, the largest example of which is now to be found on Bartlett Island (Figure 3G).

Medium-Grained Granite.

It appears that the disturbance creating these younger ring-dikes is to be associated with a somewhat smaller reservoir of granitic magma which had accumulated beneath the western part of the island (Figure 2D). In time the roof rocks above this magma were sufficiently weakened to disintegrate and settle much as they had done above the earlier magma chamber further east. This time, however, the pluglike block of roof rock was not completely isolated by the curved fractures but remained attached along its northern and western sides, tilting and flexing downward and gaping widely along its southern and eastern margins (Figure 2E). Concurrently, the lighter granitic magma was forced upward to fill the enlarging curved fissure and to solidify as a medium-grained granite. Formed in this fashion within the curved fracture, the medium-grained granite constitutes a ring-dike.

Again let us imagine ourselves looking down from high above the area. The second major subsidence occurred as fracturing took place, largely in the coarse-grained granite, along the eastern and southern sides of an incompletely isolated pluglike block (Figure 3H). As magma entered the steep curved fracture, it first advanced upward and then turned more horizontally to spread northwestward over the top of the slowly sagging block. Upon crystallizing, the ring-dike of medium-grained granite formed. At our level of observation this arc-shaped body of granite would have appeared unusually wide, if measured in the northwest-southeast direction (Figure 3I), because of the manner in which the magma spread laterally over the top of the subsiding block.

Disposition of the Rock Units.

There was little change in the rock architecture subsequent to the intrusion of the medium-grained granite; and a fair picture of how the various rock units were disposed at this time, therefore, may be gained from a study of the fold-out geologic map. The geographical distribution of each rock type is portrayed on the map by a different pattern. These distributions were determined by a careful study of many thousands of outcrops or ledges throughout the area.

We see from the map that the medium-grained granite and coarse-grained granite constitute the core of the island. Surrounding this core is a fringe of older rocks. On the west are the huge sheet of diorite and smaller patches of the Ellsworth Schist. In contact with the coarse-grained granite on the north, east, and south is the Bar Harbor Series. In the southwest lobe of Mount Desert Island is a wide band of fine-grained granite, which apparently extends eastward to Sutton Island. South of this granite is the Cranberry Island Series.

Several arc-shaped masses or ring-dikes are shown on the western side of Mount Desert Island. These appear to extend southwest to Bartlett Island where they integrate to form a single and much wider ring-dike. On the east side of Mount Desert Island near Great Head is a conspicuous but short segment of a ring-dike. As we have already seen, the body of medium-grained granite is more properly to be considered a ring-dike also.

In addition, several small patches of granite are shown on the map at the south end of Mount Desert Island (south and east of Bass Harbor), on Great Cranberry, and on Baker Island. These granites may be related to the medium-grained granite of the area, and they may represent only the uppermost exposed portions of much larger and more deeply buried bodies. At the north end of the island (two miles north of Town Hill) is a small mass of granite apparently identical with the fine-grained granite around Southwest Harbor.

Typical rock material of the shatter-zone, Ingraham Point (south of Hunters Beach). This breccia is composed of dark blocks of basaltic dike rock and lighter blocks of disrupted beds of the Bar Harbor Series held together firmly by a whitish granitic matrix. Layering (bedding) is visible in some of the lighter blocks. After formation, this chaotic assemblage was extensively cracked causing it to weather and erode in irregular fashion.

Basaltic dike (dark) cutting through the fine-grained granite (light) at Schoodic Point. The more closely spaced fractures (joints) in the dike have caused it to disintegrate rapidly and become eroded, largely by wave action, to a lower level than that of the adjacent granite. At the left, near the water, is part of another basaltic dike.

Faults.

Minor changes in the geologic structure have come about as a result of faulting, a phenomenon involving slipping or displacement of the rocks along large fractures. Commonly faults are nearly vertical fractures along which the rocks on one side have moved downward relative to those on the other. The traces of a number of small faults are shown on the geologic map around Salisbury Cove, at the north end of the island. One large fault in the western part of the island probably extends from the vicinity of Seal Cove northward under Seal Cove Pond and Hodgdon Pond (to the north) to some point past Pretty Marsh. It appears to have cut off the western extremities of the masses of coarse-grained granite and medium-grained granite, bringing these rocks abruptly into contact with the diorite along a straight north-south line (compare Figure 3I with the map).

Dikes.

Not shown on the geologic map are the numerous dikes encountered over most parts of the island. These relatively small bodies were formed when hot magma flowed into straight open fractures in solid rock and crystallized. Dikes are to be found by the hundreds on the barren mountains and along the rocky shores. Most conspicuous are the black dikes, composed of basaltic rock or diabase, that range from a small fraction of an inch to more than 50 feet in width. On the barren mountains some of the larger dikes may be seen to extend in straight courses across country for thousands of feet. Dike margins or walls are relatively smooth and parallel. Local deviations in one margin are generally matched by corresponding deviations in the opposite wall. This supports the belief that the dikes are truly fillings in opened cracks.

Most dikes exhibit extremely fine-grained margins that pass gradually into progressively coarser rock nearer the center. These fine margins are due to rapid cooling and solidification of the dike magma when it came into contact with the cold wall rock. They constitute part of the evidence that the dike material was at one time liquid and that the dike rock itself is younger than the rock with which it makes contact. That the original dike material was extremely hot is indicated by the fact that the older rock is baked and recrystallized against the dike. The width of the baked zone is usually a matter of only a few inches in the case of small dikes, but it may reach several feet in large ones.

The direction of trend has been carefully measured for about a thousand dikes over the entire island, and the results of this study have proved very instructive. Basaltic dikes in the coarse-grained granite, for example, trend a little to the west of north. Basaltic dikes in the fine-grained granite immediately to the south, however, show two distinct trends. One set trends a little to the west of north, as noted above; but the other set trends somewhat to the east of north. Many examples are known, furthermore, where northwesterly trending dikes cut through and are chilled against northeasterly trending dikes indicating the northeasterly set as the older. By further reasoning we are led to believe that the coarse-grained granite is younger than the fine-grained granite because it contains only the younger set of dikes. It is now known that there are on the island at least four basaltic dike sets of distinctly different ages.

The widespread distribution of these dikes in time as well as space suggests that the basaltic magma that formed them was derived from some extensive and perennial source or layer deep within the earth.

Light-colored dikes, generally granitic in composition and up to several hundred feet wide, are common and widespread. They are perhaps most conspicuous within the body of diorite where they contrast strongly in color with the dark enclosing rock. In the severely fractured rocks (the shatter-zone) immediately surrounding the coarse-grained granite small dikes of granite are extremely abundant. Here granitic magma was forced into cracks that developed when the huge pluglike mass of the earth's crust collapsed into the magma reservoir below.

Veins.

The rocks of Mount Desert Island are cut by multitudes of small mineral-filled crevices known as veins. They indicate cracking of the enclosing rock and subsequent filling of the fissures with mineral matter. These veins range up to a few feet thick and appear as swarms of parallel or somewhat braided ribbons traversing the bedrock surface. Two general trends are common throughout the area, north-south and east-west.

In the road-cut at the north end of Somes Sound (just east of the bridge over the tidal inlet) a myriad of white quartz (veins cuts medium-grained granite. Near the old site of the Champlain Monument on Sea Cliff Drive (one mile southwest of Seal Harbor village) is perhaps the largest quartz (vein in the region. It is approximately three feet wide and, contrary to popular belief, does not extend "clear across the island."

Many of the veins carry, in addition to quartz, small amounts of epidote, hematite, or other minerals. They resemble somewhat those veins from which gold and other valuable minerals are obtained; but, fortunately, they appear to be of no economic value. Where feldspar is an abundant constituent and the texture is coarse-grained, the veins are called pegmatites. Such veins are common in the zone of shattering immediately surrounding the coarse-grained granite.

Parallel fractures (joints) in the coarse-grained granite of Western Point. Locally joints are highly regular and closely spaced. Elsewhere few or only incipient fractures are visible. Generally one set of fracture directions dominate, but locally two intersecting sets may be seen.

33665

3. SCULPTURING OF THE LAND FORMS

Creation of Mount Desert Range.

Thus far we have considered the first two phases of our geologic story, namely the accumulation and alteration of the stratified rocks and the intrusion of molten rock matter into these units. From the time of intrusion of the medium-grained granite to the Pleistocene or glacial period, an interval of possibly 350 million years, we have a very incomplete record. It is quite likely that sediments were deposited in this area, but no direct evidence of their existence has yet been found. The area was probably above sea level for extended periods, during which erosion was the active geologic process. And so here we may begin the third phase of our story, the sculpturing of our present land forms.

Weathering and erosion had been so extensive that by Cretaceous time (about 135 million years ago) the land for hundreds of miles around had been reduced to a low nearly featureless plain (peneplain). Only an occasional low mountain relieved the monotony of the landscape. Such prominences are known as monadnocks and are generally composed of relatively durable rock. Erosion in the vicinity of Mount Desert Island had stripped away a thick cover of stratified rocks and laid bare the numerous masses of intrusive rock (Figure 2F). The tenaceous character of the coarse-grained granite was in striking contrast to that of all neighboring rock types, and the granite effectively resisted by the very existence of the Mount Desert Range, which may be considered a true monadnock.

By Tertiary time (about 60 million years ago) the mountain range was a nearly continuous ridge trending roughly east and west. The crest line was relatively straight, and small streams with essentially parallel trends descended one flank of the range to the north and the other flank to the south. All slopes were relatively gentle and there were no through-going valleys such as seen today. Mount Desert Island and the numerous neighboring islands were at this time all joined and part of the mainland.

During the Tertiary Period the land was uplifted and slightly tilted toward the sea. This movement reactivated the streams throughout the whole region, and they began to excavate their valley bottoms by sweeping away the disintegrated rock material and carrying it far out to sea. As a result the old peneplain surface was recut into low hills and shallow valleys. But the Mount Desert Range was little changed.

Subsidence which followed brought the shoreline many miles to the northwest, roughly to its present position. As the ocean inundated this newly dissected region, it flooded the valleys or converted them to bays and estuaries; it transformed ridges into headlands, and it isolated hills and prominences to form islands. A highly irregular shoreline was thus created — one perhaps not unlike that of today. Although this was a most significant change in the relative position of sea level in this region in late geologic time, it was by no means the last. The shoreline has shifted many times since then, sometimes rising, sometimes falling. Many of these fluctuations accompanied and followed glaciation in a complicated sequence.

Work of the Glacier.

A thick layer of ice began to build up over most of eastern and southern Canada in the Pleistocene Epoch (about one million years ago) and spread slowly southward across New England. At its margin this glacial sheet was probably somewhat thinner; but, farther back from the advancing front, the ice layer may have attained a thickness of many thousands of feet. Roughly 20,000 years ago the ice front reached the Maine coastal region, having traversed with little difficulty the relatively smooth peneplain to the north. The only real obstacle encountered in this region was the Mount Desert Range, which extended transverse to the direction of ice movement. Gradually the ice sheet piled up on the north flank of the mountains and the flow was deflected around the east and west ends of the range. The ice gradually ascended the northern slope of the range and then spilled through the low places in the divide to form long fingers, which descended the small valleys on the south flank. The streaming of glacial ice across the divide carved the spillways into deep saddles and increased the size and erosive power of the finger-like protrusions. As the saddles deepened, rather straight, through-going valleys were cut completely across the range. The long east-west mountain range was finally dissected into a series of detached peaks by the north-south trending glacial troughs.

As viewed from the north or south, these troughs now present a U-shaped cross section with steep walls and broad, flat floor. This form is in striking contrast to the V-shape so typical of nonglaciated stream valleys. So effective was abrasion and scour by the glacial ice tongues that some valleys were cut to depths below the present sea level. Somes Sound is a fine example of such a valley that has subsequently been flooded by the sea to form a fjord. In other valleys glacial scour was adequate to produce basined areas in the bedrock floor. Today these depressions are water-filled and constitute the most beautiful lakes in the park.

Eventually the glacier thickened, and the Mount Desert Range was completely buried beneath a deep layer of ice. At this time the entire area was subjected to the erosive forces of the continental glacier.

In its prolonged course across country, the glacier scraped soil from the underlying bedrock and plucked large blocks from the cracked and jointed ledges. Much of this material became entrapped by the ice and was carried southward for great distances. Frozen rather rigidly into the moving ice, it served as an abrasive to wear away the hard bedrock. Evidence of such abrasion may still be found in the polished, grooved, and striated ledges; and many boulders and cobbles exhibit the markings of reciprocal wear.

The Bubbles from south end of Jordan Pond. These hills were reshaped as glacial ice poured through the large valley toward the observer.

As the ice melted the enclosed material was gradually dropped to form an uneven deposit of a glacial debris. It blanketed the scoured bedrock surface beneath and locally upset the original drainage pattern by obstructing small streams and forming ponds and swamps. Most of the debris from the waning ice sheet was deposited as a heterogeneous mixture known as till and composed of silt, sand, and pebbles. Some, however, was washed and assorted by glacial meltwaters to form layered or bedded deposits of sand and gravel. Such deposits are well exposed today in the many gravel pits on the island. Low ridges of glacial debris built across several of the large glacial troughs obstructed drainage and raised to even higher levels some of the lakes in the scoured glacial valleys.

Unique features marking the glacial invasion are the large boulders scattered widely over the island. These are commonly several yards across, and many are composed of rock quite foreign to the area. Most striking perhaps are the large "erratics" perched precariously on pedestals or steep hill slopes. An impressive example is the huge balanced boulder on the east slope of the South Bubble at the north end of Jordan Pond.

Erosion in Progress.

We might close our story of the geology by pointing out that the dominant geologic process in action today in the area is that of erosion. Even the most resistant rock, the coarse-grained granite, is slowly disintegrating and wearing away. This is readily seen on mountain tops where frost action and other forces of weathering have pried loose small rough fragments of the granite. These have accumulated in slight depressions on the barren ledges to form small thin patches of gritty or gravelly material; or they have washed down slope to the small streams and thence been swept away to the ocean. Huge slabs and blocks of granite are continuously breaking loose and accumulating at the foot of steep mountain slopes to build up such extensive deposits of fallen rock as may be seen on the east side of the road just south of the Tarn or on the southern side of the Bubbles facing Jordan Pond.

Work of the Ocean.

Along the coast line may be seen the aggressive attack of the ocean in its attempt to wear away the land. The loose rock material on the shore is kept in more or less constant agitation by the waves. Blocks of rock dislodged from the sea cliffs are gradually rounded to boulders and pebbles or eventually ground to fine sand. This loose debris, known as beach, is a restless material. Its finer constituents may be carried away to deeper water or may drift along shore and accumulate in favorable sites to form sand bars and beaches. A few bars, such as the one extending from Bar Harbor village to Bar Island, are uncovered at low tide, thus, temporarily connecting the small and larger islands. The sands and finer materials derived from the exposed headlands may be swept to shelter in the coves and bays to help maintain the fine swimming beaches there. Where the shore line is more severely attacked by waves, existing beaches are composed almost wholly of pebbles and cobbles. During heavy storms these large stones are tossed inland to form high ridges or storm beaches well back on the shore. Several storm beaches may be seen on the south side of the island where they form steep, high sea-walls.

Another characteristic feature of the rocky shore is the sea cave, developed in a cliff face close to sea level. Caves and chasms form where extensively cracked or jointed rock is unable to withstand the weathering and wave attack. As blocks of rock become dislodged from the cliff face and are moved away by the waves, a small cave may be formed. Under favorable conditions, as at Anemone Cave and The Ovens, caverns many yards deep may develop.

At the close of the geologic drama of Mount Desert Island, perhaps one thought should stand out above all others — a thought so admirably expressed by the late eminent William Morris Davis, who wrote:

> As the waves rise and fall in broken rhythm on the shore, as the tide flows and ebbs across the littoral belt, so the seas of former times have risen and fallen in uneven measure on the uneasy land; the rocks have grown and wasted; the ice of the North has crept down and melted away; — all shifting back and forth in their cycles of change. Only one scene lies before us of the many that have floated through the past.

Graceful steps (below) lead to the new Hull's Cove Visitor Center and Headquarters of the Acadia National Park (above). Granite blocks were quarried just north of Stonington, Maine.

PART II

SELF-GUIDED TRIPS FOR MOUNT DESERT ISLAND

1. BAR HARBOR AND THE SHORE PATH

On the shore near the municipal pier is an exposure of the Bar Harbor Series. These fine-grained sedimentary rocks appear in layers or beds that incline northward. The inclination or dip of the beds is downward into the water at an angle of about 15-20 degrees from the horizontal, and the waves roll up on and off the smooth layers without undermining them. Looking down upon these exposures, the layers appear to be cracked by several sets of parallel fractures or joints. On the freshly broken surfaces the beds appear bluish gray or slightly lavender colored, and they consist principally of siltstone. They were originally deposited as horizontal layers but have since been tilted and jointed.

Walk out on the end of the pier and look northward into the northern part of Frenchman's Bay. The island to the left front is Bar Island. It is connected to Mount Desert Island by a wave-built bar over which a car may be driven at low tide. This is the bar from which Bar Harbor derives its name.

At the shoreline on Bar Island and extending about 50 feet up the cliffed slope are more beds of the Bar Harbor Series. Although not very obvious from this angle of view, these beds also incline to the north at a low angle. Overlying these beds is a thick layer of diorite that constitutes the higher parts of the island. It also slopes northward and is exposed almost continuously along the north shore of Bar Island. The northward tilt explains why the Bar Harbor Series is completely hidden except on the south side of Bar Island where the beds emerge from beneath the thick layer of diorite.

On Sheep Porcupine, the next island to your right, the structural relations are similar, but, due to your oblique line of sight, the northward inclination of the layers is more obvious. Beyond is Burnt Porcupine Island on the southern part of which certain tilted layers, rich in the mineral feldspar, have weathered nearly white and contrast with the darker beds and overlying diorite. Just beyond, on Long Porcupine Island, are steep cliffs formed principally of diorite. Only a little of the Bar Harbor Series is exposed on the extreme right, near the water's edge.

Looking east along Sand Beach. Ledge in the foreground is the coarse-grained granite. Beyond the beach is rock of the shatter-zone.

Bald Porcupine Island, formed of a great sheet of diorite, inclines to the left. In foreground beds of the Bar Harbor Series incline gently seaward along the Bar Harbor Shore Path.

You may now visualize the existence, at some earlier time, of a continuous thick sheet of diorite extending from island to island and inclining downward under the water only to emerge again, together with beds of the Bar Harbor Series, on the islands and mainland about five miles away along the north side of Frenchman's Bay. The diorite was injected as a magma between the beds of the Bar Harbor Series so as to form a flat thick sheet or sill. Later warping and folding of the rocks caused the tilted structure now seen in the islands.

The asymetrical profile of the Porcupine Islands is due to the tilted rock layers. The resistant diorite has served everywhere as a protective capping for the underlying beds, except on the south side of the islands where erosion of the soft sediments undermined the tilted diorite sill and produced a steep south slope. The southerly-moving glacier may have been influential in producing the profile. As it glided up over the gentle north side of the islands, it plucked large blocks from the south side and tended to accentuate the cliffed slopes. The role of wave action in cutting back these steep southerly slopes has probably not been important.

The sand beach near the pier is partly man-made. Take the footpath that starts just above the beach and leads southeast along the shore. At the far end of the beach, note how smooth the ledges were worn by the glacier. Glacial scratches and grooves may be seen on some surfaces.

Just before reaching the point in front of the hotel, note how the beds dip toward the northwest but a little further on appear to swing back again to a more northerly inclination. Most of these beds are gray, but a few show the pale lavender color so indicative of the Bar Harbor Series. The true color, however, is locally masked by a thin film of rusty iron oxide. Notice how the closely spaced joints give the exposures a ragged, blocky appearance. On the point in front of the cliff and near the high water line is a stack or towerlike form, several feet high and composed of undisturbed but isolated beds of siltstone.

Partly imbedded in the grass near the path are several large boulders composed of granite quite unlike any of the bedrock on the island. These erratics have been imported by the glacier and were probably dislodged from ledges at least 20 miles to the northwest.

Just beyond the hotel is an excellent spot to study the Bar Harbor Series. The gray and lavender beds are locally stained brown and jointed in many directions. Two sets of joints at right angles commonly dominate. Note how sharp corners of the broken ledge, above high water mark, gradually give way to rounded corners and smoothed surfaces below, denoting attack by waves.

Beyond is a shingle beach composed of flattened or platelike pebbles and cobbles with well-rounded corners and edges. The flat shape is characteristic of beaches formed where the Bar Harbor Series constitutes the bedrock. This is due to the fact that shinglelike pebbles are produced by abrasion of thin rock slabs broken from ledges of the well-bedded sedimentary rocks. In striking contrast, the more massive rocks (granite) break up into roughly equidimensional blocks and form beaches with more spherical pebbles.

Just beyond is another large boulder, about ten feet high, perched rather delicately on the dipping rock layers. Like the others, this boulder is composed of very coarse granite with white feldspar crystals more than an inch across. Cutting through the boulder is a fine-grained granite dike about six inches thick. Examine the smooth ledges of the Bar Harbor Series here and note the abundant small pods and veinlike streaks. These consist mostly of quartz that appears to fill parallel cracks and gashes in the rock a few inches long. Another shingle beach lies just beyond.

At the next point of land the beds in front of the stone wall dip eastward at about 30 degrees. The rather sudden change in direction and amount of inclination along this shore indicates that the sedimentary layers have been warped and tilted unevenly. Near this point are several dark colored dikes that cut through the sedimentary beds. One dike is about 25 feet thick. It was formed when a large crack opened in the Bar Harbor Series and permitted the influx of molten rock (magma). The cold sedimentary rock quickly cooled the basaltic magma and caused it to solidify by crystallizing feldspar and pyroxene.

A short distance beyond may be seen beds, rich in feldspar and weathered whitish, like those on Burnt Porcupine Island. This rock is cut by a grayish dike, about 18 inches thick and somewhat irregular in trend, which stands slightly above the adjacent ledge. Another large dike (200 yards ahead) cuts the sedimentary rocks and crosses the path at the small foot bridge a short distance beyond. About 150 feet before you reach the bridge, note the contact of this dike against the sedimentary rocks exposed near the stone wall at the edge of the path.

At the end of the path, a short distance beyond the foot bridge, look back on the shore and note the cave formed at the base of the cliff and near the highwater mark.

Directly off shore is Bald Porcupine Island and an artificial breakwater. Like the other islands nearby, Bald Porcupine has an asymmetrical profile and shows beds of the Bar Harbor Series dipping beneath a thick capping of diorite.

Weathering along fractures (joints) in the coarse-grained granite near Sand Beach. Disintegration and decomposition of the granite is proceeding most rapidly along numerous regular joints creating deep parallel crevice-like depressions in the ledge. The large furrow running at right angles to the others has also formed along a joint. It is now partially filled with small particles of granite that became detached from the ledge surface and washed into it by the rains. Grass has started to grow on this newly created soil. Newport cave area.

2. BAR HARBOR TO PARADISE HILL OVERLOOK

From Bar Harbor village drive about three miles northwestward along Route 3 toward Hulls Cove. You will pass the Bar Harbor-Yarmouth ferry landing on the right and soon come close to the water's edge at "The Bluffs" with a splendid view of Frenchman Bay. About half a mile beyond, turn sharp left at the National Park entrance. A short distance ahead turn sharp right to the new Park Headquarters, and visit the new Headquarters building where you may obtain much useful information pertaining to the area in the form of books, maps, and park programs.

As you return to your car, examine the large blocks of granite used in the balustrade at either side of the stepped walkway. This material is much coarser than any granite in the ledges of Mount Desert Island and was quarried in the town of Stonington on Deer Isle, Maine (see the section "By Boat to Isle au Haut"). The rock is characterized by numerous pale lavender-pink feldspar (potassium-rich) crystals up to two inches across. Some of these appear encased by a thin shell or coating of whitish feldspar (sodium-rich). Many of the large crystals are clustered, and some appear to have been turned so that their long dimensions are parallel to one another.

After leaving the parking area, turn right onto the new Park road and proceed to Paradise Hill overlook, at the highest point along the road, half a mile ahead. Park in the parking area along the left side of the road and walk out to the line of large stones. Well to your right, as you face the water, is the pier for the ferry to Yarmouth, Nova Scotia. Directly beyond is Bar Harbor village with its municipal pier. Unless the tide is near the high stage, you can see between the two piers the bar connecting Bar Island (the nearest island) with Mount Desert Island.

Successively to the left and more distant from Bar Island are Sheep Porcupine, Burnt Porcupine, and Long Porcupine Islands. This line of islands divides Frenchman Bay into two parts, a northern half that lies before you and a southern half to the right. Each island consists of a thick layer of dark colored diorite resting on top of beds of the Bar Harbor Series (see the "Bar Harbor and the Shore Path Section"). From here you may readily visualize the former existence of a continuous thick sheet of diorite connecting the islands across the bay and inclining downward toward the north (left) beneath the water.

About two miles to your front is a tiny isolated island known as Bald Rock. Directly in line with the rock but on the far side of the bay is Calf Island. Immediately to the left of Calf Island and more or less blending in with the mainland is Preble Island and then tiny Dram Island, just beyond which lies the village of Sorrento. These three islands are the northern counterparts of the Porcupines in that they are composed of a thick layer of diorite more or less sandwiched between layers of the Bar Harbor Series to form what is known as a sill. This diorite sill, furthermore, inclines to the south and probably continues all the way beneath the bay, emerging again in the Porcupines to your right.

A mile to your extreme left is Hulls Cove and Hulls Cove village. The high forested ground beyond the cove is held up by diorite, which overlies nearly flat beds of the Bar Harbor Series. Undoubtedly this diorite represents part of the great sheet or sill that is found in the Porcupines and islands to the north and that occupied the space directly beneath your feet just prior to the intrusion of the coarse-grained granite. It is believed that the layers of rock beneath the northern half of Frenchman Bay are relatively flat; but near the bay margins, they are bent upward so as to emerge in an incomplete ring of islands. Thus, the form of these curved layers, as they lie one above the other, resembles that of a stack of saucers.

Near the opposite shore, directly to the front and to the right of Bald Rock, is a long low island called Stave Island. To the right and at the same distance is Jordan Island (largely hidden by Long Porcupine Island). Both islands and the mainland beyond are composed of fine-grained granite like the fine-grained granite of Mount Desert Island. Its pink color shows up even at this distance in a clear afternoon light.

On the skyline a little to the left of Preble Island may be seen a group of mountains known both as the Franklin Hills and the Gouldsboro Hills though they are not in the township of either name. The asymmetrical and apparently highest peak (on the left) is Schoodic Mountain (1069 feet) about 12.5 miles away. At the right of the group are the 2 peaks of Black Mountain (1049 and 1094 feet). The apparently lower peak in the center is Tunk Mountain and is actually the highest (1157 feet) and most remote (18 miles) of the group. Except for the Mount Desert Range, this group of mountains is the highest in the eastern Maine coastal belt and is developed on a mass of granite of the same shape and size as that of the Mount Desert Range. Strikingly enough the granite itself is nearly identical with the coarse- and medium-grained varieties on Mount Desert Island. Moreover, these two separate masses appear to have been similarly formed and are in many respects unique in this part of the state. Only Schoodic Mountain appears from this vantage point to exhibit an asymmetrical form, though all the other peaks each possess one. Such asymmetry is attributable to the work of the ice sheet. As the continental glacier moved southeasterly across the range, it smoothed and broadly rounded the northwestern flanks but was able to pluck blocks of granite from the jointed rock on the southeastern sides leaving steep cliffed slopes.

Far to the left of Schoodic Mountain (about due north) is Hancock Point, and on the skyline just to the right of the point is Lead Mountain (1475 feet) 32 miles away. If you look carefully in line with this mountain, you may see a tall tower. This is not a fire tower on the mountain itself but a flashing beacon (clearly visible after dusk) on the ridge only 11 miles away.

Let us look quickly now at some of the features in the southern half of Frenchman Bay. The geology of this portion will be considered in the section on "Cadillac Mountain to Champlain Mountain Overlook."

Beyond Bar Island is Bald Porcupine Island with its artificial jetty extending about half a mile to the right. Over Bar Island and Sheep Porcupine Island you may see Ironbound Island, the largest island in the entire bay. Beyond Ironbound Island and limiting the bay on the east is the Gouldsboro Peninsula, which extends south (to the right) to Schoodic Peninsula, another section of Acadia National Park. Near the south end of the peninsula is Schoodic Head (over Bald Porcupine Island) and Schoodic Point. To the right of the point (over Bar Harbor) is Egg Rock and its lighthouse. Further right are 3 of the main mountains of the Mount Desert Range. These are from left to right: Champlain Mountain (1058 feet), with Huguenot Head lower and to right, Dorr Mountain (1270 feet), and Cadillac Mountain (1530 feet).

Your present location is on the shatter-zone. Just a few hundred feet up hill, across the road, is the coarse-grained granite enclosing only occasional fragments of the Bar Harbor Series. Better exposures of the shatter-zone will be seen further south along the road.

Looking northeast across Frenchman Bay from summit of Cadillac Mountain (telephoto lens). In the foreground are rounded ledges of the coarse-grained granite. Bald Porcupine Island (with breakwater) and Long Porcupine Island beyond show steep cliffs on their south sides. These islands are composed of the diorite and represent parts of a huge sheet that underlies most of the bay. Beyond are Stave Island and the mainland, both composed of the fine-grained granite.

3. PARADISE HILL OVERLOOK TO CADILLAC MOUNTAIN

About 0.4 mile south of the overlook is a stone bridge over Duck Brook. Cross the bridge and park in the parking area on the left of the road. Walk back to the center of the bridge and look upstream. Note the shape of the valley formed by Duck Brook. The steep slopes, which form a V-shaped cross-section for this valley, indicate that the brook is actively eroding and cutting down rapidly through the high ground (held up by massive granite) just before it passes onto the low ground (underlain by weaker rock of the shatter-zone) nearer the shore. The bridge is constructed of the medium-grained granite that comes from Hall Quarry on the west shore of Somes Sound.

Walk back past the parking area a few yards to the large road-cut in badly weather-stained rock composed of beds of the Bar Harbor Series. This part of the shatter-zone is not severely disturbed; but one can see how the rock layers have been contorted, ruptured, and pulled apart. If you continue to drive straight south along the road, you will pass through more of the shatter-zone and gradually into the main mass of coarse-grained granite. About 0.7 mile from the bridge you will cross Eagle Lake Road on the overpass. Continue on 0.8 mile to the Summit Road which turns off sharply to your left. You are now well within the largest rock unit of the island, the coarse-grained granite. Due to its resistant character, this granite makes up all the mountains and high hills on the island. The pink color of the rock contributes greatly to the aesthetic quality of the region; but this color, though conspicuous, is neither universal nor restricted to the coarse-grained granite.

As you ascend the Summit Road to the top of Cadillac Mountain, note in the many road-cuts how the granite is broken by nearly vertical fractures or joints as a result of earth movements. Note also the strong development of gently inclined fractures that have broken the granite into huge slabs or sheets resting one above another. This latter phenomenon, known as sheeting, represents a special type of jointing. As you pass from cut to cut note how the sheets are everywhere arranged nearly parallel to the land surface. When the granite formed, it crystallized at high pressure under a deep cover of rock. As this thick cover was removed by erosion, pressure on the granite was reduced; and the rock attempted to relieve the internal stresses by expanding. The easiest and principal direction for expansion was perpendicular to the land surface. As the granite cracked at right angles to the direction of expansion, thin slabs or sheets formed parallel to the land surface.

In several large road-cuts you will see one or more of the basaltic dikes so common and widespread throughout the granite. These appear as steep, dark bands several yards wide across the face of the cut. Along the road are numerous places to observe the bedrock surfaces that were rounded, smoothed, and polished by the glacier.

Park at the small turnoff, 1.3 miles up on the Summit Road, for a view of Eagle Lake, the second largest lake on the island. This lake occupies a basin in large part scoured out of the granite bedrock by the glacier. An accumulation of glacial till at the lower end has aided the impounding of water in this valley.

Park at the Summit Parking Area and walk to the lookout. Note the highly irregular coast line. This is typical of a submerged coast. To get your bearing, notice that the Summit Shop is now about west of you. Look east across Frenchman's Bay to Schoodic Head, nine miles away. Beyond may be seen a series of additional bays and promontories. To the south is the indentation of Otter Cove and the causeway over which you will drive later. Note how boldly Mount Cadillac stands out above the low, gently rolling coastal region. Its prominence, as well as that of the entire Mount Desert Range, is due to the resistance of the coarse-grained granite of which it is composed.

Walk the Summit Loop (footpath) in a counterclockwise direction, and note the rounding and smoothing of Champlain and Dorr Mountains (to the east) by the ice sheet. Near the northern part of the loop is a 5-foot boulder of a coarse granite like those along the Shore Path in Bar Harbor. Examine it and see how strikingly it differs from the granite upon which it rests. This glacial erratic was carried by the ice at least 20 miles from some ledge to the northwest. Six feet west of the boulder the coarse-grained bedrock still preserves its glacial polish.

On the north side of the road at the parking lot exit (where the North Ridge Trail begins) is a small deposit of gravellike material. Examine it closely and note it consists of rough, somewhat weathered particles of quartz and feldspar. This is a residual soil formed by the mechanical breakdown or disintegration of the coarse-grained granite. This material is continually forming on the granite surface and, if not trapped in slight depressions or retained by vegetation, is washed down the mountain slope.

Overleaf: Looking south over Jordan Pond to the Cranberry Islands Glacial ice scoured the valley between the Bubbles (lower left) and Jordan Ridge (right center) and formed the depression for Jordan Pond.

4. CADILLAC MOUNTAIN TO
CHAMPLAIN MOUNTAIN OVERLOOK

Descend the Summit Road and at its lower end turn right. At the road junction about half a mile ahead bear right toward Ocean Drive. Within the next 2.7 miles this road will take you around the north end of Cadillac Mountain, across a broad flat valley floor, known as Great Meadow, to an underpass beneath the Bar Harbor — Otter Creek Road (Route 3).

Just before reaching the underpass you will observe a road on the right to Sieur de Monts Spring. A brief visit to this locality (a few hundred yards away) will provide an opportunity to sample the cool spring water, visit the Abbe Museum of Stone Age antiquities, and study the displays (some geological) in the Park Nature Center.

Return to Ocean Drive and proceed through the underpass. Beyond the road curves left past Bear Brook Picnic Area and then swings back and begins to ascend the north flank of Champlain Mountain. About half a mile from the underpass at the highest point along the road is Champlain Mountain Overlook. The parking area here, on the seaward side of the road, is 253 feet above the ocean and offers a fine view of the southern half of Frenchman Bay.

Near the shore a little to the left is the Thrumcap, a very small island. To the right, two miles offshore, is Egg Rock with its lighthouse. Farther to the left is Ironbound Island, the largest island in Frenchman Bay and composed almost entirely of diorite. It is presumably a southern extension of the huge diorite sheet underlying the northern half of Frenchman Bay (see the trip "Bar Harbor to Paradise Hill Overlook"). Here in the south half of the bay, however, the diorite sheet is very flat and at one time probably extended at least all the way to Great Head (on the extreme right) on Mount Desert Island. Its thickness was perhaps only a few hundred feet. Supposedly much of the diorite encountered in the shatter-zone along the east coast of Mount Desert Island had its origin in this extensive sheet, which was thoroughly disrupted here at the time the coarse-grained granite formed.

Beyond and to the right of Ironbound is the Gouldsboro Peninsula with its shores of pink fine-grained granite precisely like the fine-grained granite of Mount Desert Island and probably continuous with it under the southern half of the Bay. The southern part of this peninsula is known as Schoodic Peninsula, a large part of which lies within

Acadia National Park. The conspicuous flat-topped hill (440 feet) appearing just over Egg Rock is Schoodic Head. South of the Head the skyline drops abruptly to a low, flat promontory at the end of which is Schoodic Point. Just north of the point are the towers of the Naval Radio Station.

To the left of Ironbound Island and beyond, the pink fine-grained granite can be seen for miles along the shore of the islands and the mainland. Still further to the left, several of the Porcupine Islands are visible, and in the distance one may see the Franklin Hills (Gouldsboro Hills). The geology of these hills and islands is described in the visits to "Bar Harbor and The Shore Path" and "Bar Harbor to Paradise Hill Overlook."

You are now near the contact between the coarse-grained granite and the breccia of the shatter-zone. Near the south end of the roadcut is the granite. It is finer grained than usual because the granite magma cooled quickly near the contact. Note also the color of this rock, due to an abnormal amount of dark-colored mineral. Locally small inclusions, probably of the Bar Harbor Series, are abundant within the granite.

Looking south across Frenchman Bay to the mountains of Mount Desert Island from U. S. Route #1 near Sullivan. The highest peak to the left of center is Cadillac Mountain.

To the north, within the granite, is a huge block of the Bar Harbor Series with beds of gray, green, white, and pale lavender siltstone or quartzite. This mass is locally cut by veins and dikes of granite material. Further to the north is much fine-grained diorite as large blocks in granite and cut by granitic veins. Here also a basaltic dike three feet thick cuts through the breccia.

Looking southeast from Anemone Cave to the open ocean at relatively low tide.

5. OCEAN DRIVE

Anemone Cave.

Drive on, about 1.5 miles to the south, along the base of the steep cliffed slopes of Champlain Mountain to the overlook parking area on the left. Follow the path down to the shore, and walk over the rocks to Anemone Cave. You are now in the ancient zone of fragmentation (the shatter-zone). The ledges along the shore are composed of breccia. Note how the diorite and Bar Harbor Series were thoroughly shattered, impregnated with granitic material, and restored to a consolidated and coherent rock. More recently this mass of breccia was somewhat cracked and jointed by disturbances far less intensive than those which formed the old zone of fragmentation. Note how the ledges tend to be more severely cracked in certain places than in others. This selective fracturing prepared the way for the formation of Anemone Cave. The rock material that originally occupied the space within the cave was more intensely cracked than that now seen forming the cave roof. Frost action and the work of waves were largely responsible for the gradual excavation of this large opening. If the tide is not too high, one may enter the cave. Notice how extensive, nearly horizontal fractures in the roof have controlled the falling of large slabs and the formation of the broad, flat cave roof.

You may look into the beautiful clear tidal pools on the irregular floor of the cave, but do not disturb them. Here live the sea anemones, pink algae, various types of snails, and many other marine forms.

Sand Beach.

Continue along the main road about 0.6 mile toward Sand Beach. As you drive into Sand Beach Parking Area, note how Sand Beach has been built across a small bay so as to dam the stream and form a long, narrow pond. Walk east along the beach, and notice that the level of the fresh water pond behind the beach is slightly higher than that of the ocean. At times there is no apparent outlet for the small pond, and the fresh water drains through and beneath the beach to the ocean. During periods of relatively low tide you may see it escaping in tiny rivulets rising out of the lower part of the beach and flowing in small rills down to the shoreline. At other times a small stream flows from the pond to the ocean.

Examine a bit of the beach sand. It is unusual in that it contains a large percentage of shell fragments in addition to the more ordinary grains of quartz and feldspar. Tiny fragments of clam and mussel shells and green spines of sea urchins are readily detected in this material.

Now, standing near the top of the beach and facing the ocean, note the rocky coast line that forms the west (on your right) side of the little bay. These ledges of coarse-grained granite extend southward a mile and a half past Thunder Hole and Otter Cliffs to Otter Point. On the left side (east) of the bay there is little granite, and the ledges are composed mostly of the older country rock (Bar Harbor Series and diorite) into which the granite magma intruded. The contact between granite and older rock must lie, therefore, somewhere beneath Sand Beach.

Examine the ledges at the east end of the beach. From a distance a distinct layering may be visible, and at one place beneath the gently inclined layers a small cave has formed. Closer inspection reveals that most of this material consists of various sized fragments of the older rocks imbedded in a matrix of granitic material much like that to be seen later near Hunter's Beach Head Parking Area. Here, however, a much greater variety of fragments may be seen. Fragments range from angular to round in shape and from light to dark in color. Some show the characteristic layering, and many have developed one or more zones or concentric layers of slightly different color or texture in their marginal portions. The texture of these fragments has been made slightly coarser by the heat and metamorphic action of the granite. This locality is also within the shatter-zone and these ledges are evidence of the intense shattering of the old country rock that accompanied the emplacement of the large mass of coarse-grained granite.

Return to the parking area and look east across the small pond. The long ridge on the skyline in front of you is held up by the coarse-grained granite. This granite forms a long, narrow, dikelike body wholly within the shattered country rock. The body extends north and south for nearly half a mile. It has already been described as a short segment of a ring-dike.

Note the steep round hill (The Beehive) immediately to the northwest. The steep slopes and cliffs of this prominence, as well as of those further north, are controlled largely by the nearly vertical fractures or joints so strikingly displayed in this mass of granite. Note also the set of nearly horizontal joints that aid in the formation of stepped slopes. These stepped cliffs came about as a result of quarrying by glacial ice.

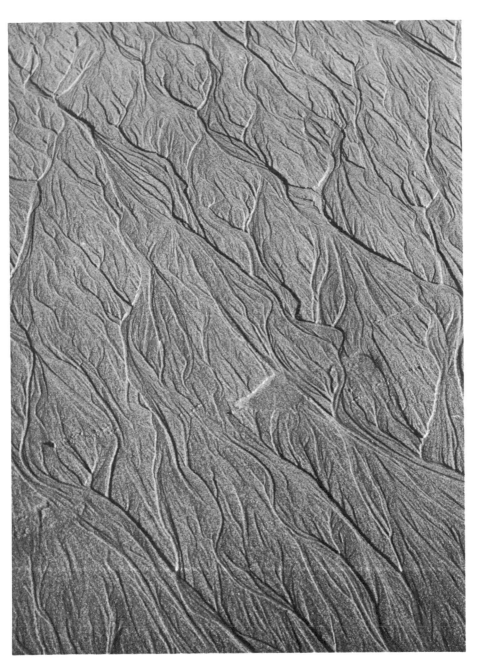

Rill marks formed on Sand Beach by water running down the beach toward the observer. Note footprint at right of center.

Great Head.

A footpath leads about one-half mile to Great Head, one of the highest headlands on our Atlantic Coast. At the top you are 145 feet above the sea, and here you will have a superb view of the eastern shore of Mount Desert Island and the southern part of Frenchman's Bay. To the south is Otter Point. Six miles to the east, across the bay, is Schoodic Head. Two and a half miles to the northeast is Egg Rock and Egg Rock Lighthouse. To the north, along the eastern part of Mount Desert Island, is a relatively low narrow belt of land bounded on the west by abrupt slopes of the Beehive and Champlain Mountain. This low area lies mostly within the zone of intense shattering formed at the contact of the granite and older country rock. Beautifully exposures of this highly disrupted zone may be seen all along the shore from Great Head nearly to Bar Harbor village. Look north along the shore and note two shallow caves formed in the ragged coastal ledges. Around the first large bend in the shore line is Anemone Cave; just beyond, Schooner Head forms the last feature visible on the shore.

Near the top of Great Head the bedrock consists mostly of diorite, somewhat cut across by thin light-colored veins of quartz and feldspar. As you walk back to the west along the trail, note how the ledges change from massive diorite to a light-colored granitic rock filled with fragments of diorite and the Bar Harbor Series. Note the long low ridge a few hundred yards ahead (to the west) composed of the coarse-grained granite. This is the ring-dike visible from Sand Beach.

Thunder Hole.

About 0.7 mile further south along the main road from Sand Beach Parking Area is Thunder Hole. This chasm, developed along large joints in the coarse-grained granite, gets its name from the noise emitted as the waves roll in and slap against the crevice walls, compressing the air trapped in the cavernous end. The chasm is still slowly being enlarged as granite blocks and slabs are dislodged by frost action, wave impact, and other processes. The detached fragments tumble to the base of the crevice and are ground to pieces or swept out to sea by the action of the waves. Thus, the waves prevent the mouth of the chasm from becoming choked with fallen debris. By their continued agitation, the large rounded boulders in the bottom of Thunder Hole (barely visible at low water) keep grinding away the floor and sides of the chasm.

Great Head, formed of diorite and rocks of the shatter-zone.

Monument Cove.

Drive to the Monument Cove Parking Area just ahead. Walk to the shore path, and look down into the small cove flanked by steep granite cliffs. Observe the conspicuous steep joints and the extensive sheeting fractures that incline seaward at a low angle. On the north side of the cove is a small stack of granite about 20 feet high and separated by joints from the adjacent cliff. The coarseness of the beach material below is unique in the Park. Here is a fine example of a boulder beach. When heavy waves strike the shore, the rolling and grinding of the well-rounded boulders becomes audible.

Otter Cliff.

Your next stop is at the Otter Cliff Parking Area, one-half mile ahead, just before you reach the divided highway. Go to the sidewalk at the head of the cliff and look out across the water. About seven miles to the front (nearly east), across the southern part of Frenchman Bay, is Schoodic Head, a flat-topped hill (440 feet high) located at the south end of Schoodic Peninsula. This hill and much of the adjacent lowland, a detached portion of Acadia National Park, is underlain by the fine-grained granite. Four miles to the northeast of you is Egg Rock Lighthouse. Egg Rock consists of ledges of diorite like that of the Porcupine Islands. A mile and a half to the north are the cliffs of Great Head, the top of which is 145 feet above the sea. Immediately to the left, at the head of Newport Cove, is Sand Beach. Between your present position and the beach may be seen the rugged shores of coarse-grained granite. The jagged, blocky outline is due to abundant fractures or joints, which so thoroughly dissect the granite.

Your present position is about 110 feet above the ocean, and the rocks forming the high cliff are also composed of the coarse-grained granite. Down the path a short distance, in the face of the steep slope, is a large mass of very fine-grained rock completely enclosed by the granite. The numerous tiny feldspar crystals in this rock suggest it is of igneous origin. It may represent a large fragment of a felsite dike or volcanic material.

Stack and boulder beach, Monument Cove. The rounded boulders were once angular-shaped like the granite blocks in the cliff.

Otter Point.

Drive on 0.3 mile and park in the Otter Point Parking Area. Walk down to the shore in front of the center of the parking area. You are now close to the shatter-zone again. Here is a mass, more than 100 feet long, of fine-grained sedimentary material completely isolated in the granite. The rock shows a bedded structure with pale lavender, white, green, and gray layers commonly less than an inch thick. It represents a huge fragment of the Bar Harbor Series, which dropped into the granite magma and was recrystallized to a tough brittle quartzite by the heat of the enclosing granite. Note the numerous small veins composed mostly of quartz and feldspar, which cut across the different colored beds.

Face the ocean and then walk to the right about 75 yards. Here are numerous large boulders, ranging up to 10 feet through and composed of the coarse-grained granite. They have been washed out of the glacial drift, which can still be seen in the steep bank just back from the shore. Note that the granite of these boulders is slightly coarser than that of the ledges upon which the boulders rest. This indicates that the boulders are true erratics and have undoubtedly been brought from ledges nearer the center of the coarse-grained granite mass where the original magma crystallized to a coarser texture.

Notice the tiny rounded inclusions (small bodies) of fine-grained, darker colored rock in many of the boulders and in the underlying ledges as well. These little rock fragments seem to be widespread throughout the coarse-grained granite body. Except for these inclusions the granite here and for some distance ahead is practically free of fragments. You are now within the main body of coarse-grained granite. Note the systematic arrangement of joints, which dissect the granite and form a step-like surface. The edges and corners of many steps have been rounded by the glacier and, to a lesser extent, by the ocean. The preservation of glacial markings (striations) on many of these surfaces indicates little marine abrasion since glacial time.

Reverse your direction along the shore, and walk back about 150 yards to a point roughly opposite the east entrance to the parking area. Here you will observe a mass of the Bar Harbor Series, about 300 feet long, forming the coastal ledges and cut by numerous dikes and veins of granitic material. You are now back in the shatter-zone, but here the old country rocks have not been so thoroughly broken up. You will see relatively little breccia. Instead, the originally flat beds have been flexed and bent down toward the granite. Note that the layers incline at a low angle to the northwest (landward), which is directly toward the body of coarse-grained granite. These beds were bent and dragged down from their horizontal position as the roof rocks above the large chamber of granite magma sagged and collapsed into the hot molten material.

Little Hunter's Beach.

After leaving the parking area continue along the main road 0.8 mile to the causeway and bridge across Otter Cove. From the causeway look northward to see Cadillac Mountain separated from Dorr Mountain by a deeply cut (glacier-worn) notch.

Drive on to Little Hunter's Beach, 1.6 miles ahead. After leaving the causeway you will pass through a long road-cut in coarse-grained granite essentially free of incorporated fragments of older rock. As you drive on, note how the amount of granite in the road-cuts decreases and masses of older rock gradually become predominant. Upon reaching Little Hunter's Beach you will have passed from the main body of coarse-grained granite back into the shatter-zone.

Park in the small area at the end of the stone bridge, and take the path to the cobble beach below. Notice how round the cobbles have been worn by the waves. If the sea is rough, you may be able to hear these large stones grinding against one another as they are rolled to and fro. Walk out on the ledges from the far end of the beach to see an outstanding example of the material making up the shatter-zone. Note the variety displayed by the fragments of diorite and Bar Harbor Series firmly embedded in a granite-like matrix. The origin of this striking rock (breccia) and the formation of the shatter-zone are explained in the next section.

Hunter's Beach Head.

Continue driving along the road about a quarter of a mile to the roadside parking strip at Hunter's Beach Head. Park close to the curb on the ocean side of the road. Here you are about 100 feet above the ocean. Note the deep bend or angle cut back into the cliff face below you. This has been controlled by extensive fractures or joints in the cliff.

At this station you are located within an ancient zone of shattering (shatter-zone) near the contact with the coarse-grained granite. Accompanying intrusion of the large body of granite magma into the solid country rock composed mostly of the Bar Harbor Series and sills of diorite, intense fracturing and granulation took place. The diorite and

The shatter-zone, Hunter's Beach Head. Light colored granite dikes and veins cut through and disrupt dark colored diorite forming an angular breccia.

sedimentary beds were severely ruptured and broken into blocks and fragments of various sizes. Much of this rubble adjacent to the magma became incorporated in the granite melt; farther away the broken rock was impregnated with exhalations from the magma or cut across by a myriad of granitic dikes and veins. Still more remote from the granite contact, the older country rocks were less intensely broken and only rarely cut across by protrusions of granite. As the granite cooled and crystallized, all breaks in this ancient shatter-zone were sealed and the rubble-like material made a solid whole (see geologic map). Rocks of this type are known as breccia and are well illustrated in the long road-cut here. Examine the cut and note the large blocks of fine-grained dark diorite, cut across by dikes of relatively fine-grained granite and coarse veins of pegmatite. In places the diorite appears to have been shattered and included as smaller angular fragments in the granitic material. To the northwest (your left as you face the road-cut) the diorite gives way to very fine-grained white, tan, and gray-colored rocks of the Bar Harbor Series, opposite the end of the paved sidewalk.

6. JORDAN POND ROAD TO BAR HARBOR

From Hunter's Beach Head drive on 2.5 miles, beneath two granite underpasses, and bear right at the junction with Jordan Pond Road. About 0.7 mile beyond the junction you will pass Jordan Pond House on the left. About 200 yards beyond the house, turn left to the parking area and walk down the short foot path to Jordan Pond. This long body of water lies in a steep-sided valley carved largely by glacial ice. From the shore is an excellent view (worth photographing) of the broad steep-sided valley through which the glacial ice flowed. Somewhat obstructing the ice movement in the center of the valley were the two rounded hills beyond the pond known as The Bubbles. Although Jordan Pond fills an ice-scoured basin, its level has been raised considerably by a thick deposit of glacial debris that blocks the valley near here.

Continue northward along the main road for nearly three miles to Bubble Pond. On the way you will climb high along the western slope of Pemetic Mountain, gaining a fine view of Jordan Pond and The Bubbles. The huge boulder perched precariously on the east side of South Bubble (nearest you) is composed of very coarse-grained granite not found in any of the ledges on the island. It was carried by the glacier from some ledge at least 20 miles to the northwest.

Nearly three miles north along the road from Jordan Pond parking area, turn right into the parking area at Bubble Pond. As does Jordan Pond, Bubble Pond rests on the floor of a glacier-worn, U-shaped valley formed between Cadillac Mountain on the left and Pemetic Mountain on the right.

About 0.7 mile along the main road from Bubble Pond is a small roadside parking strip. From here is an excellent view across Eagle Lake to Sargent Mountain. Near the south end of the lake, below Sargent Mountain, are the Bubbles. In this view they appear as two somewhat long and slender hills smoothly rounded on the north and west sides but ragged and cliffed on the south and east. During the glacial period the ice moved southward over these hills smoothing off the north and west sides and plucking large granite blocks from the south and east sides. Across the road opposite the parking strip are fine examples of ledges that have been smoothed and polished by the glacier.

By driving northward 1.5 miles from the parking strip, you may return to the Eagle Lake Road overpass. From here it is roughly another 1.5 miles to the village of Bar Harbor.

7. EAGLE LAKE ROAD TO SOMES SOUND

Follow the Eagle Lake Road (Route 233) west, out of Bar Harbor, to the granite overpass about a mile from the village. Drive under the overpass and park at the small turnout on the right about 150 feet beyond. Walk back and examine the deep road-cuts, but stay well off the pavement. Note that steep joints in the long cut run in a number of different directions. Sheeting is well developed with irregular to curved horizontal fractures from a few inches to a foot apart. More striking perhaps are the many long, thin inclusions of medium to dark gray rock (commonly up to a foot long and one to two inches thick) turned with their longest dimensions horizontal. The proximity of the main contact of the coarse-grained granite is indicated here by the slightly finer-grained texture of the granite.

Drive on one mile to Eagle Lake and park in the lot on the right side of the road. Walk down to the lake and notice the deeply notched profile of the Mount Desert Range. Looking up the lake you will see the steep-sloped outline of Pemetic Mountain. To the right is South Bubble, and on its eastern slope, clearly silhouetted against the sky, is the huge boulder. Next right is North Bubble. To the left of Pemetic Mountain is the flank of Cadillac Mountain. A more expansive view of the mountains may be obtained by walking to your left (as you face the water) along the carriage road. From here, first Penobscot and then Sargent Mountains come into view on the extreme right.

Notice along the shore a nearly continuous line or ridge of granite boulders, which have accumulated in response to ice shove. As the lake ice crowds landward in the colder months, it shoves along boulders partly frozen into it in the shallow region along the shore. Over a period of many years boulders may be moved great distances, only to accumulate in long welts or ridges, called ice ramparts, at or close to the shore line. If you walk as far as the bridge (300 yards) over the outlet to Eagle Lake, you will pass several tiny coves each more or less isolated from the open lake by a rampart. On the shore at the lake outlet, one may see that the water level of Eagle Lake has been raised about 5 feet by an artificial dam. As you walk back along the road, notice how the lake has been formed by a natural deposit of glacial material which constitutes a rather continuous broad ridge a short distance back from the lake. Formerly, when the lake was about 10 feet higher than at present, it overflowed at the lowest point along this natural dam to form a spillway; and the deep valley seen at the bridge was cut as water washed away the unconsolidated glacial debris and exposed the granite ledge beneath.

Drive west one-half mile from the parking area to the large open working in glacial material on the right side of the road. An examination will reveal that this material differs from that in most of the sand and gravel pits on the island. The deposit, though somewhat bedded, is extremely poorly sorted; that is, pebbles and cobbles of various sizes are mixed with sand, silt, and fine clay particles. The pebbles and cobbles, furthermore, are poorly rounded as a rule; and the deposit must have built up rather rapidly and under conditions quite different from those that formed the nicely bedded and sorted materials observed in most other large pits. Evidence of extensive abrasion and wear, during movement by the glacial ice, is displayed by the smoothed, faceted, and striated surfaces on many of the cobbles in this working.

View of Pemetic Mountain and adjacent hills from north end of Eagle Lake. Glacial ice moved across the area away from the observer. The line of boulders near shore represents an ice rampart.

Further along the road is a good view of Sargent Mountain, on the left; and, across Aunt Betty Pond in the distance, are Western Mountain and Beech Mountain with its fire tower. About 0.3 mile from the gravel pit you pass onto a narrow belt of recrystallized granite, numerous exposures of which appear on the right side (north) of the road. About 1.1 miles from the large gravel pit you will pass Norway Road leading off on the right. Half a mile further along the main road is the most recently opened quarry in the coarse-grained granite. Like all others, however, this quarry is now abandoned. Roughly a mile beyond you will leave the area of coarse-grained granite and pass over to the younger mass (a ring-dike) of medium-grained granite. Just ahead Route 233 joins Route 198 near the north end of Somes Sound.

8. SARGENT DRIVE, SEAL HARBOR AND THE TARN

At the junction of Route 233 turn left (south) onto Route 198 and notice the medium-grained granite in the deep road-cuts a short distance ahead. Here you can see the effect of jointing upon the color of the granite. The massive portions of the rock are unweathered and light-colored. In the highly jointed and sheared portions, however, weathering has produced a pink-colored rock by oxidation of the iron.

One mile south bear right onto Sargent Drive (Route 3) and check your mileage carefully at this junction. Almost exactly 1.4 miles beyond is a small turnoff, at the foot of a steep slope on the east (left) side of the road, just large enough to accommodate two cars. Do not stop here but continue on slightly down hill 0.1 mile to a larger area on the left, where a small brook (which may be dry during certain seasons) crosses under the road. This location offers an excellent view of Somes Sound, which is a fjord, and an opportunity to take photographs. Here at about its narrowest part, the Sound appears to cut through the Mt. Desert Range, passing between the steep slopes of Norumbega Mountain on your side of the water and Acadia Mountain directly across from you. This narrow valley through the mountains constricted the flowing ice and forced it to cut deeper here than elsewhere. Soundings near here show the water to be as much as 168 feet deep. Across the Sound, to the right of Acadia Mountain, is Hall Quarry. This most

The Tarn from its north shore. The U-shaped valley between Huguenot Head (left) and Dorr Mountain (right) was formed by glacial erosion.

productive quarry site on the island supplied granite of fine quality for construction of the Philadelphia Mint.

Your present location, at the little brook, is close to the contact between the two principal granite bodies of the island. In the road-cuts north of the brook is medium-grained granite; south (ahead) is coarse-grained granite. The former rock is not only finer textured but carries less dark-colored mineral than the latter.

Walk 250 yards beyond the turnout to a large bronze marker fastened to the steep face of the road-cut in the coarse-grained granite. Just beyond the marker several basaltic dikes are exposed in the high cliffs. One of these has been split apart and intruded by another dike of light-pink felsitic rock.

Continue driving on Route 3 past the golf course and through Northeast Harbor. As you pass into the village and out on the other side, notice the rocks in the road-cuts. You will pass successively out of the granite, into the old shatter-zone, and then back into the granite.

Turn right at the next road junction and drive on to Seal Harbor. A little more than two miles from this junction you will pass Bracy Cove where a seawall of pebbles and cobbles has been built by the waves. Long Pond, the small body of water on the left, was formed when an old bar was built across the bay at the point where the highway now passes. This is an excellent spot from which to view or photograph the Bubbles and Penobscot Mountain. Note the balanced boulder on South Bubble.

Beyond is an excellent sandy beach at the head of Seal Harbor. Follow Route 3 through the village toward Bar Harbor. You will soon pass the small village of Otter Creek and enter the broad U-shaped valley between Dorr Mountain on the left and Champlain Mountain on the right. As you pass, notice the great piles of talus or fallen rock that are accumulating at the base of the steep slopes. To the left of the road is the Tarn, a small shallow pond formed on the floor of this broad ice-scoured valley.

Glacial erratic boulder on South Bubble. Eagle Lake in distance to left and Gouldsboro Hills near center on skyline.

9. SOMESVILLE TO SEAWALL PICNIC AREA

The Medium-Grained Granite.

From the bridge in Somesville follow Route 102 due south for 1.6 miles to the road that turns left to Hall Quarry. Here, at the north end of Echo Lake, is a deep road-cut in a finer-grained phase of the medium-grained granite. This rock constitutes the youngest ring-dike of the area and affords an excellent opportunity to observe the color effects of weathering. From a distance, bands of altered pink granite several yards wide contrast sharply with the fresh gray rock. From closer inspection it is clear that fresh granite becomes highly discolored along individual fractures in the rock and this alteration effect commonly extends back an inch or so on either side of a visible fracture. Where fractures are numerous and closely spaced, the whole rock appears reddened. The pigment causing the discoloration is produced by oxidation (rusting) of finely dispersed iron in the feldspar of the rock.

Continue on Route 102 for 0.7 mile and, after passing close to the shore of Echo Lake, note the road-cuts on the left are in medium-grained granite. About 0.3 mile ahead, a second road turns left to Hall Quarry. If time permits you may wish to visit the little village and inspect the large excavations from which a good quality of stone has been taken for many decades. Permission to enter the quarries may be obtained by inquiring at Hall Quarry village.

Echo Lake Beach.

Continue on Route 102, past long road cuts in the coarse-grained granite, to the road leading to Echo Lake Beach (1.3 miles from the second Hall Quarry road). Turn sharply right onto this road and drive 0.6 mile to the parking area. Walk down to the beach. The lake, which is almost 2 miles long and one-third mile wide, lies in a broad flat-bottomed valley, enlarged and greatly deepened by glacial scour. During Pleistocene time a huge tongue of glacial ice poured through this north-south valley (toward you) abrading and wearing away the solid granite. In time the valley was greatly broadened and deepened, perhaps several hundred feet. Later, when the glacier had thickened to form a continuous sheet that buried the mountain tops, the direction of ice flow was more southeasterly and oblique to the trend of the valley. This oblique motion allowed the ice to glide up the right wall of the valley and transform it to a more gentle and smoothly rounded slope. Along the left valley wall, however, the ice plucked blocks of jointed granite and carried them away leaving the much steeper and more ragged slope known as Beech Cliff. Note in these cliffs how the combination of vertical joints and horizontal sheeting fractures has aided glacial erosion by preparing rectangular granite blocks for the plucking action. You are observing here a feature characteristic of Mount Desert Island, valleys with steep ragged western slopes and gentle rounded eastern slopes.

The beach here is not native; the sand and fine gravel have been imported to form a thick layer over the original, silty to muddy beach material. Human attempts to improve upon nature in this fashion are not always successful; but, judging from the nature of the long narrow termination of Echo Lake here, wave and current action should be favorable to the preservation of this artificially prepared beach.

Long Pond (Great Pond).

Return to the main road and continue south along Route 102 for 1.4 miles to the Seal Cove Road. Turn right and follow Seal Cove Road for 0.6 mile; then bear right along the main road one mile to the pumping station at the south end of Long Pond. The steep slope down which you drove to the pumping station is formed on a thick deposit of glacial material, which accumulated near the south end of the great valley in which Long Pond now lies. This deposit constitutes a gigantic natural dam across the valley, which permitted the lake to rise until it spilled over at its north end and drained northward through Somesville and into Somes Sound. The lake is 4.0 miles long and averages half a mile wide. At its constriction between Mansell Mountain on the left and Beech Mountain on the right, the lake is only one-eighth of a mile wide. Just as we saw at Echo Lake Beach, the glacial ice crowded through between the two mountains here cutting down the original saddle between them and forming a deep through-going valley across the Mount Desert Range. Here also is to be noted the steeper, more cliffed slope of Mansell Mountain on the left as compared with the slope of Beech Mountain on the right. Once again the asymmetry of valley shape is attributed to the oblique movement of the glacial ice.

Seal Cove Pond and Oak Hill (Bald Mountain).

Drive back along the main road about 0.7 mile and turn sharp right on the gravel road to Mill Field, Western Mountain, and Oak Hill.

On some maps Oak Hill is indicated as Bald Mountain. Check the mileage on your car. The road at first passes through a huge gravel pit and for a few hundred yards may be difficult to locate. One's efforts in searching out the way here, however, will be well rewarded because ahead lies over 3 miles of what many may consider the most beautiful and relaxing drive on Mount Desert Island. Before leaving the gravel pit notice the thickness and cross-bedded nature of the deposit. This pit gives one an idea of the nature of the material which blocks off the south end of Long Pond.

Just beyond the gravel pit a narrow but well-graded and well-maintained road begins. It winds over low flat ground and comes close to the contact with the coarse-grained granite. Then it turns southward along the base of the steep slope of Western Mountain and finally swings westward to Seal Cove Pond. The road wends its way through beautiful clear woods of spruce, birch, and maple with large patches of cedar in the more swampy areas. About 3.3 miles from the entrance to the gravel pit, a turn-off on the left leads to Oak Hill. Do not turn here, but continue 0.7 mile to Seal Cove Pond, and note that the woods soon changes to one of maple and birch. Park at the end of the road and walk to the shore. The low hills across the pond are held up by diorite, and beneath your feet is the same rock, slightly south of its contact with coarse-grained granite. The pond trends north-south and follows a fault for over a mile (see geologic map).

Drive back 0.7 mile and turn right toward Oak Hill. Half a mile further along turn sharply right and drive around the loop at Oak Hill parking area. Park and walk west a few hundred feet to the top of Oak Hill. Approximately northeast of you is Bernard Mountain, the nearest and highest peak of Western Mountain. Look directly opposite from the mountain and locate the deep notch in the salt-water shoreline. This notch is at the head of Seal Cove (to be visited later) and represents the deep narrow valley cut by the small stream that now drains Seal Cove Pond at the foot of Oak Hill. On the skyline above the notch, and extending for some distance to the right and left, is Deer Island and other smaller islands of Penobscot Bay. To the right of the notch, on the horizon, may be seen the Camden Hills, nearly 40 miles away. On the extreme right is Blue Hill with its pronounced terracelike southern slope. The mountainlike island on the horizon, just left of the notch, is Isle au Haut, about 20 miles away. Left of Isle au Haut, but at a lesser range, is a long group of islands consisting principally of Swans Island and Long Island. At your extreme left Baker Island (part of Acadia National Park) may be visible, and just to the right of this the Duck Islands may be seen in clear weather.

In your immediate vicinity is rough, knobby topography so typical of ground underlain by the diorite. This poorly jointed, irregularly broken, dark gray rock crops out here through a rather thin soil cover. The highly weathered nature of the rock is indicated by the greatly rust-stained surfaces and the broken and disintegrating blocks and outcrops. Much of the rock is so altered that it crumbles under the slightest pressure. Extensive disintegration of this diorite has led to the formation here of a coarse, granular, residual soil cover, quite different from the glacially transported soil found elsewhere. Examine the soil material at the base of a sloping ledge surface and note how coarse, sharp, and irregular the rusty particles are. These are merely tiny fragments derived from the adjacent diorite ledge by both chemical and mechanical weathering. The rusty color of the soil and rock surface stains is the result of chemical attack (weathering) by the elements (air and water) on the iron-bearing minerals so abundant in the diorite. The sparsely scattered vegetation and somewhat scrubby growth of juniper and sumac are indications of a low fertility soil probably derived by decomposition and disintegration of diorite.

Seawall Picnic Area.

Retrace your way to Route 102 and turn south (right) through the village of Southwest Harbor. About 0.6 mile south of the center of the village, turn left from Route 102 and drive on Route 102A for 3.0 miles to Seawall Picnic Area. Just before reaching the area you will pass Seawall Pond on the right. This pond formed when the low ground back from the shore was dammed by the high storm beach along which the main road runs for a short distance here. Do not stop here (you will return soon) but drive on toward the campground about 0.3 mile ahead.

Turn left between the two stone posts opposite the entrance to Seawall Campground, and drive to the shore. The large ledges along the shore represent the Cranberry Island Series and consist mostly of volcanic tuff, a rock formed by the accumulation and consolidation of dust, fine ash, and rock fragments produced by explosive volcanic eruptions. Before consolidation, much of this volcanic material was reworked by the waves and currents of a very ancient ocean. Examine the ledges or a number of the loose blocks or cobbles on the beach. Notice that they consist of fine-grained, thoroughly hard material and that within them are what were at one time abundant fragments of various types of older rock. Most of these old fragments are angular and small, but some are more than five feet long. They represent pieces of shattered lava flows, fine-grained sediments, and even older volcanic tuffs which have been mixed with fine volcanic ash. The whole mass has been subsequently consolidated.

Face the water and then follow the shore around to your right. This will take you around the small point of land and into the broad bay-like indentation in the shoreline. A firm level path runs along the very top of the beach. Note that the loose material upon which you are walking is composed mostly of the volcanic tuff and that the pebbles and cobbles are rather angular in form. After walking about 200 yards along the beach, notice how much more rounded the beach pebbles and cobbles have become. Most of the pebbles and cobbles here were originally angular fragments broken from the ledges on the point of land from which you have just walked. These fragments have been tossed to and fro by the waves on the beach and have gradually drifted along the shore in the same direction in which you have been walking. Since the degree of rounding depends upon the amount of wear, those beach materials which have traveled the farthest (those where you now stand) will be more rounded than those nearer the source at the point. Here the cobble beach appears as a level curved ridge forming a barrier across the low ground to the right. A little further on you will reach the highest part of the beach, composed of extremely well-rounded cobbles and impounding behind it a small body of water. The pond water drains slowly through the somewhat permeable beach deposit and at low tide can be seen trickling out of the sand and gravel and running down the lower part of the beach. This entire cobble beach was constructed by the waves, and during severe storms cobbles may be rolled or tossed over its crest. Storm beaches built in this wall-like form are known as seawalls.

Return to the main road and park just south of Seawall Pond (about 0.3 mile north of the campground entrance). On the shore opposite Seawall Pond is an extensive flat ledge forming a low promontory and flanked on either side by a cobblestone beach. This ledge consists of a fine-grained sugary granite. Viewed from a distance it appears to be composed of flat sheets separated by numerous and extensive horizontal fractures or joints. Closer inspection shows the ledge to be dissected by two sets of well-developed vertical fractures (joints) arranged at right angles. These joints in combination with the horizontal set cause the ledge to break down into thin rectangular blocks.

The light-colored granite receives its pink cast from finely scattered grains of iron oxide (hematite), which are locally so distributed as to produce a mottled or speckled effect. Note within the granite the abundant pods and irregular veins of milky white quartz up to several feet long. At the vein margins may be seen numerous large crystals of a feldspar called microcline. This mineral is usually white or buff, but locally it exhibits a beautiful green color and is then referred to as amazonite. Though none of this material is of gem quality, many of the better crystals have been broken out or damaged by thoughtless visitors. Many of these veins are true pegmatites due to their very coarse texture. Feldspar crystals one or two inches across are common, and black mica (biotite) crystals half an inch across may be found.

Walk south (to your right as you face the water) about 250 yards until you come to three black dikes running parallel to each other but at a slight angle to the shore. The largest dike is about eight feet wide. The others range between one and three feet and may be seen to merge into a single dike toward the north.

Examine the rock into which these dikes have been intruded, and note that it is not the fine sugary granite you have just examined to the north but the volcanic tuff observed further south. Retrace your steps for a short distance to the north to find the contact between the granite and volcanic rocks. At this contact several thin veins of granitic material may be seen extending from the granite into the adjacent tuff. The linear arrangement of fragments in the tuff roughly parallels the shore, but the contact between granite and volcanic material (seen near high tide level) trends roughly at right angles to the shore and cuts cleanly across the structure of the volcanic rocks.

Now carefully trace the black dikes back toward the granite and note how they terminate at the granite contact. This is most striking in the largest dike, if it is not covered by the tide. These relations lead us to the conclusion that the volcanic tuff was intruded by the dikes and is, therefore, older. The granite probably formed later than the dikes and is the youngest rock here.

10. SEAWALL PICNIC AREA TO SEAL COVE

Return to the car and continue along the main road toward Bass Harbor. About 1.3 miles west of the campground entrance you will reach the Tremont-Southwest Harbor town line and a parking area on the left (seaward) side of the road. The small body of water before you is Ship Harbor, and this location merits further exploration. The harbor occupies a low area developed on soft volcanic rocks like those seen at the Seawall Picnic Area. The harbor entrance is extremely narrow and confined between high, steep slopes formed on fine pink granite.

Follow the path from the parking area which leads around the east (your left as you face the water) side of the harbor. This will bring you to the high barren ledges which form the long constricted harbor entrance. At low tide the water in this narrows is too shallow even for small boats to enter. The granite upon which you are standing is part of the same body observed at the Seawall Picnic Area. It extends westward from here for nearly a mile to Bass Harbor Head, which you will visit next. Examine the rock closely and note how much finer it is than the other granites you have seen on the island. Note also that by comparison it contains mostly quartz and feldspar with only a trace of dark-colored minerals.

Looking north from Adams Bridge across Bass Harbor Marsh to Western Mountain. The rounded peak at right is Mansell Mountain; the long peak to the left is Bernard Mountain.

Return to the parking area and continue along Route 102A for 0.8 mile. Turn left on the side road and park one mile beyond at Bass Harbor Head. This is the most southern point on Mount Desert Island, and the lighthouse here is said to be one of the most frequently photographed on our Atlantic coast. A mile out to sea is Great Gott Island and to the right of it is Placentia Island. To your right front, beyond Placentia Island, is Swans Island. To the left front the Duck Islands may be visible roughly 6 miles away.

Return to the main road and continue on through the village of Bass Harbor. About half a mile beyond the village center bear right at the road junction. Just ahead is the main junction with Route 102 and a good road-cut on the right. The rock exposed here is felsite, an extremely fine-grained material of the same composition as granite. It is considered to represent part of the Cranberry Island Series, and it probably formed as a lava flow or mass of magma which crystallized rapidly at a relatively shallow depth.

Turn left at the road junction onto Route 102. About 0.3 mile ahead is the bridge over Marshall Brook, at the head of Bass Harbor; and on the right is Bass Harbor Marsh, a deep tidal marsh at the head of the estuary. Near the bridge and further along the road are numerous outcrops of the felsite. Two miles beyond the bridge you will come to the head of Duck Cove. Here you leave the felsite and pass over onto the fine-grained granite. Roughly half a mile beyond Duck Cove, near the top of the hill, is a cross-road and a large cemetery on the left. If time permits, turn right here to observe the diorite in outcrops a short distance along the road. A special feature of the rock here is the marked parallelism of long slender crystals of feldspar. So perfect is the parallel orientation in some places that a fair cleavage is imparted to this rock.

Half a mile further along Route 102, near the head of Goose Cove, are road-cuts where you may examine the fine-grained granite. Although the crystals of this rock are of about the same size as those of the granite at Ship Harbor, this granite is somewhat less sugary or granular in texture.

About 1.5 miles ahead the route turns sharply left to the bridge over Seal Cove Brook. This brook, which drains Seal Cove Pond (not visible on the right), is deeply cut into a narrow steep-walled canyon in the diorite. Downstream from the bridge (to your left) the deep valley gradually broadens into Seal Cove. Cross the bridge and bear left onto Cape Road and drive half a mile to the shore of Seal Cove.

From the north shore of Seal Cove here, you can look westward across Blue Hill Bay to Deer Island and the Brooklin Peninsula. Along the water's edge here are fine exposures of the Ellsworth Schist. Originally this rock was composed of light-colored sandy layers alternating with darker shaly beds. Although the beds generally incline landward at about 40°, they have been intensely folded and crinkled. Abundant white quartz veins up to an inch thick run parallel to the bedding. The original sedimentary materials here have been transformed or metamorphosed to mica schist. A few beds of dark massive rock up to a foot thick may be seen here. The proximity of intrusive igneous rocks is suggested by the development of a black mica (biotite) in place of green chlorite in these schists. Face the water and then turn left and walk along the shore. You will soon observe a breccia composed of schist fragments enclosed in a granitic matrix. This is part of the great shatter-zone (see geologic map) which formed during the intrustion of the medium- and coarse-grained granites.

11. SEAL COVE TO SOMESVILLE

Along the road from Seal Cove back to Route 102 are outcrops of the granite cutting and enclosing blocks of the older dark-colored diorite. Turn left at the junction with Route 102 and drive toward Pretty Marsh Picnic Area, 3.2 miles ahead. A short distance from the road junction is another outcrop of granite and diorite, and outcrops of the dark diorite may be seen further along the road. You will soon come to the high ground overlooking Seal Cove Pond on the right and Western Mountain (Bernard Mtn.) beyond. The coarse-grained granite that holds up the mountain comes against the diorite along the line (fault) that extends the full length of the pond. The abrupt change in level from the high granite mountain to the lowland of diorite around the pond points out the great difference in resistance to erosion that these two rock types offer.

Ahead, near the Pretty Marsh Picnic Area, you will pass outcrops of granite much like that at Hall Quarry. Keep right at the road junction a short distance ahead, and you will soon pass several small roadcuts in a finer-grained phase of the medium-grained granite. Examine this rock and notice that it is even finer-grained than the material observed at the road junction near the north end of Echo Lake.

Continue along the main road to Somesville, which lies about 3.5 miles ahead.

12. BAR HARBOR TO THOMPSON ISLAND PICNIC AREA

The Ovens.

Drive northwestward along Route 3 from Bar Harbor to Hulls Cove (2.5 miles). About 1.3 miles beyond the cove, after the road completes a broad swing to the left, turn right onto the Sand Point Road. Follow this road 0.8 mile to a point where it makes a sharp left turn, and park well off on the right side. Take the steep path which leads down to the shore. It is advisable to visit this locality during the low- or medium-tide stage. At high tide many of the features along the shore here are inaccessible.

Here one sees that the Ovens are domelike or arched caves developed in the face of a vertical cliff that ranges up to 35 feet high. Each cave or oven appears to have been formed where the cliff rocks were most intensely cracked and subject to disintegration. The breakdown of the cliff face has been accomplished in part by wave action but perhaps largely by weathering. Wave action has been particularly effective in removing the loose rock as fast as it accumulated on the floor at cave mouths. Cave floors are flat and all are approximately at high-tide level. The ovens range up to 10 or 12 feet high and 10 to 15 feet deep. One is about 30 feet deep. Remember that the process of disintegration is still in operation here along the cliff face and within the caves. *Be on the alert for loosened material that may fall spontaneously or upon the slightest provocation.*

The material holding up the cliff here is a very compact flintlike rock. It is badly stained by iron (rust) along multitudes of small cracks, but the fresh material is light gray with a bluish to greenish tinge. In reality it is composed of volcanic ash which was interlayered with more typical beds of the Bar Harbor Series. In time the thick ash deposit was hardened to a volcanic tuff.

Near the highest part of the cliff and some of the largest ovens is a tall narrow arch through which you may walk. Faint layering (bedding) may be seen in the steep walls of the arch. Near the floor of the arch, the light gray volcanic tuff grades into and becomes interlayered with beds of the typical Bar Harbor Series. Bedding in these lower rocks is roughly horizontal and parallel to that in the tuffs above them. Within the tuff are numerous oval to lens-shaped solid masses up to ten inches long and oriented parallel to the rock layering. Many of these have partly weathered out of the rock leaving spongy cavernous masses.

Looking south along Somes Sound through the Mount Desert Range. Glacial ice moving away from the observer poured through the valley between Acadia Mountain (right) and Normubega Mountain (left) and carved a passageway well below seawall (a fjord).

Toward the southeast (on your right as you face the ocean) about 100 feet beyond the last oven, are the well-bedded rocks of the Bar Harbor Series. For the most part the beds are like those along the shore path in Bar Harbor village, but some beds are nearly black and somewhat slaty. Two conspicuous layers here, one 10 inches thick and a slightly higher one 40 inches thick, are composed of the same volcanic tuff as that in which the ovens and tall arch have been formed. This greenish gray rock grades vertically into the more normal beds of the Bar Harbor Series.

The abrupt change from conspicuously bedded rocks here to more massive tuff a few rods to the northwest is due to a series of faults or extensive planes of slipping. The most conspicuous faults, seen in the cliff face, are represented by steep fractures on one side of which the rock layers have been displaced downward relative to those on the other side. Disruption of the bedded rocks may be detected in the lower face of the cliff by noting the sharp termination and offset of individual layers.

Note that the ovens are not formed in the strongly bedded rocks of the Bar Harbor Series. They are confined to the more massive volcanic tuff along this shore. It is believed that the earth forces that deformed the rocks have produced strikingly different effects upon the more massive tuffs as compared with the well-bedded rocks. The former were brittle and shattered under the stress; the latter were more flexible. Shattering thus prepared the tuff for more rapid disintegration. But shattering was not of uniform intensity of extensity. Consequently, disintegration of the cliff was most rapid at sites of pronounced fracturing.

As you walk back along the shore, note how intense shattering along certain large steep joints (truly faults) has resulted in the formation of huge crevices, some of which extend for many feet back into the cliff face. Actually the narrow arch through which you may walk represents such an opening developed along a shattered zone running nearly parallel to the shore. Note that this cavity runs at right angles to the direction in which wave action might be expected to excavate. This feature lends support to the belief that disintegration by weathering is the effective mechanism, whereas wave action serves to clear away the waste which would otherwise, in time, slow down the weathering process.

Hadley Point.

Return to the car and continue along Sand Point Road for 0.6 mile till it brings you back to Route 3. Turn right onto Route 3 and drive 2.1 miles to the top of the hill where a road, leading to Hadley Point, turns off on the right. Follow this road 0.7 mile to the point on the shore of Mount Desert Narrows. Across the bay is Lamoine and Lamoine State Park. Hadley Point is formed by a triangular beach deposit one corner of which projects outward to deeper water. It probably developed as a pointed bar which enclosed a small basin. As the basin became filled with beach material, a triangular point of land was developed.

The beach here is of fine gravel and sand and is composed largely of rock material derived from the Ellsworth Schist. Face the water and then turn right and walk some 20 to 25 yards. At the water's edge you will come upon a large outcrop of light greenish gray schist composed of quartz, chlorite, and muscovite mica. Originally this rock represented beds of sandy and clay-rich (shaly) material. Later it was metamorphosed and made schistose (cleavable) as a consequence of thorough deformation and recrystallization. This property to split or cleave into thin layers is known as schistosity and is characteristic of most rocks of the Ellsworth Schist. It accounts for the thin or tabular shape of so many pebbles in the beach material here. Some idea of the severity of deformation suffered by these rocks is shown by their highly crinkled and folded appearance. Along the folded and crinkled layers thin discontinuous veins and lenses of quartz, up to an inch thick, have developed.

Thompson Island Picnic Area.

Return to Route 3 and turn right. Three miles ahead Routes 102 and 198 join Route 3 from the left. Half a mile ahead, short of the bridge leading to the mainland, turn right and park in the Thompson Island Picnic Area. Walk to the shore and observe the fine exposures of light greenish gray gneiss. Locally the rock shows a rather good cleavage and may more properly be called schist. In any event all this material belongs to the Ellsworth Schist. It differs from that at Hadley Point in that it is composed largely of white feldspar and greenish chlorite and originally consisted of beds of volcanic ash and tuff. As did the Ellsworth Schist elsewhere, however, it developed through metamorphism, a tendency to split into layers (or schistosity) which here incline at about 25° to the southeast. The quartz veins in these outcrops, unlike those in the schist at Hadley Point, are thicker, more irregular, and cut across the rock layers.

13. THOMPSON ISLAND PICNIC AREA TO BEECH CLIFF

Drive back along Route 3 and bear right onto Routes 102 and 198. About 1.5 miles ahead you will cross over from the Ellsworth Schist to the diorite. At Town Hill, 0.7 mile beyond, four roads join at a large triangle. Continue on the main route (102 and 198) and you will soon obtain a fine view of the deeply notched Mount Desert Range. About 1.2 miles from the road junction at Town Hill you leave the diorite and pass onto the ring-dike of medium-grained granite. Just beyond is a large gravel pit on the left. At the road junction ahead Route 198 turns off to the left. Here is a good spot to park and examine the medium-grained granite. The rock differs from most of that elsewhere in that it has been intensely and rather thoroughly altered, giving rise to the deep pink color. Close inspection will reveal severely fractured rock that readily disintegrates to small fragments. The dull appearance of the mineral grains on newly broken surfaces distinguishes this material at once from the fresh granite found elsewhere. This type of material is encountered along Route 198, across the north end of Somes Sound, all the way to Route 233.

Walk downhill along Route 198 to the bridge. About 200 yards uphill beyond the bridge is a deep road-cut. Here the altered and fractured granite is cut by a myriad of parallel to branching veins of white quartz. These veins testify as to the extensive deformation (in part faulting) suffered by the granite here, and the thoroughly fractured nature and altered character of the rock offer further testimony. It seems very probable that much of the granite in the immediate area of Somes Sounds was so affected, and this may explain why the rock here was so weak and readily eroded to form an ocean-flooded valley.

Return to the car and continue on through Somesville. At the bridge in the village, look left across Somes Sound to notice the broad, gentle north and south slopes of Sargent Mountain. How different this peak appears when viewed from different directions. About 0.3 mile south of the bridge turn right toward Pretty Marsh, and nearly 0.3 mile beyond turn sharp left onto the Beech Hill Road. Follow this 3.0 miles to the end. On the way you will rise to high open ground with a grand view on either side of the broad ridge (Beech Hill). At 2.5 miles you pass from the medium-grained to the coarse-grained granite, which holds up the high ground ahead.

Park in the parking area, walk across the road, and follow the short trail (about 300 yards) to Beech Cliff. From the path along the edge of the cliff you may gain a magnificent panoramic view. Looking down upon Echo Lake you will see how deeply the valley has been cut by glacial ice, completely across the formerly continuous ridge of the Mount Desert Range. The steepness of the west valley wall as compared with that of the east wall across the lake is exaggerated from this vantage point.

14. BEECH CLIFF, PRETTY MARSH AND TOWN HILL

Return to the car, drive back (3 miles) to the main road, and turn left toward Pretty Marsh. At 1.4 miles you will pass the north end of Long Pond. About 2.0 miles beyond keep left at the road junction onto Route 102. About 0.4 mile beyond the road junction is the entrance to Pretty Marsh Picnic Area. Just 0.1 mile further is a well-maintained gravel side road on the left (east) that provides a splendid drive as it winds through a varied wooded terrain to the west shore of Long Pond and back to Route 102, about a mile south of here. The round-trip distance is approximately five miles, and the drive is well worth your time.

After making the circuit, turn into the Pretty Marsh Picnic Area and park at the end of the road. Walk down to the shore. To your front is Pretty Marsh Harbor, and a mile beyond is Bartlett Island. Turn left and walk south along the shore a short distance until you come to a large outcrop of very coarse-grained rock. This material is called pegmatite due to its very coarse crystalline texture, but it differs from most pegmatite in that its composition is dioritic rather than granitic. It is composed principally of large rectangular crystals of blue-gray plagioclase feldspar and black grains of pyroxene and hornblende. It represents a small mass of the coarsest phase of the diorite.

Return to the car, drive back out to the main road, and turn left. At the junction 0.4 mile beyond, bear left along Route 102. About 0.3 mile from the road junction bear left again for a brief visit to Bartlett Island Landing at the end of the road one mile ahead. On the way you may see much granite representing large dikes cutting through the great belt of diorite here. From near the landing you may look westward across Bartlett Narrows to Bartlett Island, half a mile away. The near side of the island is composed of granite which constitutes part of a large ring-dike. The far side consists of Ellsworth Schist much like that at Hadley Point and Seal Cove.

Drive back to the main road and turn left on Route 102. This takes you past Indian Point to Town Hill, six miles ahead. On the way you will travel over diorite, and numerous road-cuts in this material may be seen. Just before reaching Town Hill your route will join Route 198. Turn right at this junction and continue to Town Hill (0.3 mile) where two secondary roads turn off on the left. Take the second road (almost opposite the fire station) which leads through low country around the north end of the coarse-grained granite mass. About 1.5 miles from Town Hill you will pass Fresh Meadow on the left. Just beyond is a private dump, the point of origin of The Great Fire of 1947. Keep straight at the next two road junctions. About 4.5 miles from Town Hill the road passes through a large crushed-rock quarry. The material obtained here is largely from the tough, flintlike beds of the Bar Harbor Series. Due to the intense fracturing and weathering, however, it may not be readily recognized. About 0.3 mile beyond the quarry is Hulls Cove Village and Route 3. Turn right here to Bar Harbor about three miles ahead.

15. THE CRANBERRY ISLES AND BAKER ISLAND

Those who have ample time in the Mount Desert area and wish to observe some interesting geology in an accessible but rather remote area should explore the shores of the Cranberry Isle and Baker Island. Boat transportation at a reasonable charge may be secured (in summer) at the pier in Southwest Harbor. On weekdays the two Cranberry Isles are also serviced by the mail boat from the same pier.

Rocks of the Cranberry Island Series, described in part in the trip to "Seawall Picnic Area," are most strikingly exposed along the shores of the Cranberry Isles. The boat will probably take you ashore at the town landing at the north end of Great Cranberry Island. From here you should walk westward (to your left as you face the water) around Spurling Point and then south as far as time allows. Ledges are plentiful and the variety of rock types is great all the way to Rice Point, the southernmost point of the island (three miles from the landing). In this distance you will pass over volcanic tuff and breccia, felsite, and spherulitic (in small, hard, round masses) rhyolite. Near Rice Point a variety of distinctly bedded rocks appear. There are white beds rich in quartz and feldspar, green beds rich in epidote, lavendar beds rich in biotite (black or dark green) mica, reddish-brown beds rich in garnet, and abundant almost black beds rich in hornblende.

If you go all the way to Rice Point, look for large blocks and boulders of a pink granite, which have been torn loose from ledges covered by the coarse beach material and the ocean. Continue east along the straight shore line for another half mile, and then turn inland up the slope to the road on Bulger Hill. Turn left onto the main road beyond, which leads back along high ground to the town landing.

The small body of water on the right of the road back is known as The Pool. It is very shallow and at low tide is not much more than a mud flat. The highly irregular shoreline on the east side of Great Cranberry Island is due to a number of tombolos (bars), composed of sand and gravel and built up by waves to integrate small outlying islands and rocks with the main island.

Granite block seawall, southeast shore of Baker Island. The blocks and slabs were broken from nearby ledges of sheeted granite along the water's edge.

The boat landing on Little Cranberry Island is on the west side, at the little village of Islesford where the Islesford Historical Museum (part of Acadia National Park) is located. Check at Park Headquarters in Hull's Cove for the Museum's open hours. From the landing walk north (to your right as you face the water) for half to three-quarters of a mile to see the fine exposures of rhyolite lavas. If you walk southward from the landing, around the southwest tip of the island, you will see an extensive and broadly curving beach running eastward for 1.5 miles to the old Coast Guard station on Bar Point. About halfway along this beach you will come to the Seawall, a high storm beach built up like a wall in front of low swampy ground to the north. Around Bar Point, at the southeast tip of the island, are numerous basaltic and felisitic dikes running through the volcanic tuffs and breccias. At low water much of the shore appears underlain by dark gray volcanic beds rich in hornblende and originally probably of basaltic composition.

From Bar Point it is nearly a mile to Baker Island, the southwestern half of which is now part of Acadia National Park. At low tide it is possible to walk to Baker Island, but the attempt should not be made without first carefully calculating the period of low water and the time required for the round trip. It would be better to make a special boat trip to Baker Island in order to have ample time for exploration there. The island furnishes a splendid view of the Mount Desert Range. It is about half a mile in diameter and rises in a single hill to 92 feet above sea level. It is composed entirely of a medium-grained pink granite, save for a northeast-trending basaltic dike 80-140 feet wide and exposed on the southwest and northeast sides of the island. Granite is almost continuously exposed along the shore and has disintegrated to large blocks. These originally were bounded by steep joint planes and gently inclined sheeting surfaces but now exhibit rounded edges and corners due to weathering and extensive wear received from the waves of the open ocean. Along the southeast shore of Baker Island may be seen a most unusual, striking seawall composed of great blocks of granite hurled shoreward by mighty waves during winter storms.

Looking northwest from near Great Head across the pond above Sand Beach to The Beehive (telephoto lens). The blocky nature of this granite hill is due to intersection of horizontal fractures (sheeting) and steep joints. As the glacier moved over this hill, more or less from right to left, it ground away and smoothed the right slope but plucked large blocks from the left slope leaving it steep and irregular.

PART III

A VISIT TO SCHOODIC PENINSULA

1. INTRODUCTION

One of the most scenic drives along the Maine coast is from Bar Harbor to Schoodic Point. The route, as you can see on a highway map, is all on good roads and takes you around three sides of Frenchman Bay, permitting you to view Mount Desert Island from a number of directions. The total round trip covers about 100 miles; thus it is advisable to make an early start for you will want the full day. This trip is so arranged that you will be spending several hours during the midday at Schoodic Point. You should go there prepared to enjoy a picnic lunch on the gently sloping ledges of pink granite while you watch the surging waves. And don't forget to pack a few crusts of bread for the seemingly famished seagulls that flock boldly wherever food is in evidence.

The Schoodic Peninsula constitutes a large neck of land, extending south for 10 miles from near West Gouldsboro on U. S. Route 1. About half of the peninsula lying south of Winter Harbor and Birch Harbor villages falls within Acadia National Park. Topographically this part of the peninsula is low and flat with a single prominence, Schoodic Head, at the south end rising to 440 feet above the ocean. A paved auto road, roughly peripheral to the area, is the principal means of access. A short branch leads past the Naval Radio Station to Schoodic Point parking area, the southernmost tip of the peninsula. Although this tip of the peninsula is commonly referred to as Schoodic Point, it might be pointed out that officially the name "Schoodic Point" refers to the south tip of Little Moose Island half a mile to the east. Nevertheless, the common usage of the name will be followed here. A gravel road, quite negotiable, leads to a parking area and lookout on Schoodic Head. This itinerary directs you over these roads and will acquaint you with the principal points of interest. Further access to this heavily wooded area is very limited by the lack of trails and improved roads.

The bedrock of the entire peninsula is composed almost entirely of one rock type. Consequently no geologic map of the area, which of necessity would be very monotonous, is provided here. The rock is fine-grained granite and is correlated with that of Mount Desert Island between Southwest Harbor village and Goose Cove. Its counterpart is

Two basaltic dikes, each about three feet wide, cutting through the fine-grained granite at Schoodic Point.

also found on Isle au Haut, another portion of Acadia National Park, where it constitutes the central part of the island. Wherever this granite is encountered along the Maine coast, it is characteristically cut through or intruded by a great number of dark basaltic dikes. These younger features are most beautifully displayed near Schoodic Point, and considerable attention is devoted to them in this itinerary.

2. BAR HARBOR TO SCHOODIC POINT

Follow Route 3 nearly to Ellsworth. Three miles north of the bridge where you leave Mount Desert Island, look right, across the valley, to the large light-colored scars on the green hillside. These are gravel pits in the thick extensive glacial deposits that cover most of the surrounding area.

Just before reaching Ellsworth turn sharp right onto U. S. Route 1 toward Machias. About 7 miles ahead you pass through Hancock village, and shortly you cross the Hancock-Sullivan bridge. Stop 1.5 miles beyond the bridge at a scenic turnout near a service station and general store. From here you may obtain the most beautiful view of the northern half of Frenchman Bay and Mount Desert Island of any along this route. You may wish to photograph this in the morning light and again in the late afternoon upon your return. The bay immediately in front is Sullivan Harbor. Its right side is formed by Crabtree Neck, its left by Waukeag Neck. From here you are looking nearly south and roughly across the length of the Mount Desert Range. This view brings out the deeply notched, steep-sided character of the channeled valleys through which glacial ice poured during Pleistocene time. From left to right the visible peaks of the range are: Champlain Mountain, Huguenot Head, Dorr Mountain, Cadillac Mountain, Pemetic Mountain, The Bubbles, Penobscot Mountain, and Sargent Mountain.

About 1.7 miles further along Route 1 is a small but pleasant picnic ground on the south side of the road. After passing the point where Route 183 turns off on the left (north), there are several spots along Route 1 that afford excellent views of Frenchman Bay and the mountains. Roughly 4 miles ahead turn right onto Route 186 near West

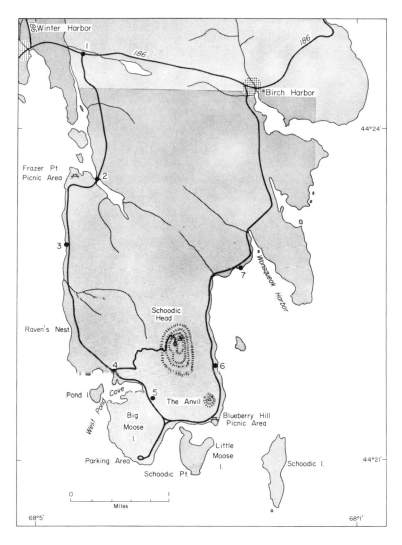

Gouldsboro. On the left, 1.5 miles from Route 1, is a gravel pit in glacial material at the top of the hill; and a mile beyond is the fine-grained granite that will be seen intermittently all the way to Schoodic Point. Three miles from Route 1 the granite is seen, cut through by a black basaltic dike. Similar dikes may be studied in detail at Schoodic Point. Three miles ahead you join the Winter Harbor road at the head of Winter Harbor. Winter Harbor village is to the right. Turn left at this junction, and 0.6 mile further on leave Route 186, turning right to Schoodic Point. You are now at (1) on the Schoodic Head map.

About 1.5 miles from this last junction you enter the National Park and cross Mosquito Harbor bridge (point 2). On the right, 0.1 mile beyond the bridge, is a turnoff to Frazer Point Picnic Area. About 0.9 miles beyond the bridge stop at the turnout (point 3) from where you may obtain an end-on view of the Mount Desert Range across the southern half of Frenchman Bay. Cadillac Mountain, the highest peak, is nearly masked by Dorr and Champlain Mountains which are slightly nearer. How gentle the slopes appear from this direction. To your left front is Turtle Island, and to your front arc two small islands of the pink fine-grained granite. To your right front is Ironbound Island with extensive cliffs composed of the great sheet of diorite already described in the trip to "Champlain Mountain Overlook." To the right of this island the pink granite appears again on Grindstone Neck.

Drive on 1.5 miles to the side road on the left (point 4) leading nearly to the top of Schoodic Head. Drive in on this road and park at the end about 40 feet below the top of the hill. The summit, which is 440 feet above sea level, may be reached easily from here by a short footpath. On a clear day magnificent views may be obtained of the Maine coast from a number of vantage points around the top. The ledges along the path are composed of the same fine-grained granite as you will soon see at Schoodic Point.

Return to the main road, park, and walk to the shore. Roughly to your front is Pond Island, composed of the pink fine-grained granite cut through by basaltic dikes. At low water this island is connected to Big Moose Island (with the radio towers) and one can walk there. On your left is a high bar composed of rounded pebbles and cobbles, which extends nearly to Pond Island. At the end of the bar is a deep channel which helps to drain and fill the inner portion of West Pond Cove (behind the bar) during successive changes in level of the tide.

Face the water, then turn right and walk along the shore a short distance. You will pass ledges of very fine-grained granite rich in inclusions (small bodies of different material) like those described later as we go on from "Schoodic Point to Bar Harbor." The inclusions may comprise 50 to 70 percent of the rock. Note that they carry small crystals of light-colored feldspar up to half an inch long and numerous black spots and knots up to three-fourths of an inch across.

Drive along the road 0.4 mile to the head of West Pond Cove. On your left just beyond you will see a low swampy area (point 5) separating the high ground to the northeast, which culminates in Schoodic Head, from that to the southwest, which is known as Big Moose Island (with the radio tower). Though this may not appear to be an island now, the swampy area was once covered with salt water. More recently, however, this area has been denied access to the ocean in part by the formation of beach deposits and the construction of the road. As the low area is gradually filled, Big Moose Island will become more completely joined to the mainland.

Bear right at the junction ahead, drive past the entrance to the Naval Radio Station, and park in the Schoodic Point parking area.

3. SCHOODIC POINT PARKING AREA

Walk to the line of large stones marking the edge of the lower parking lot. Look to the west, your right as you face the water, and see Mount Desert Island about six miles away across Frenchman Bay. In a clear light one can see Cadillac Mountain, a conspicuous, smoothly rounded elevation of barren pink granite, and the steep cliffs of Champlain Mountain immediately below. Looking slightly to the left, the island becomes very low and flat and nearly merges with the Cranberry Islands just beyond. To the left of these is tiny Baker Island and finally the Duck Islands 15 miles away.

The vast ledges here are composed of fine-grained granite like that making up most of the Schoodic Peninsula. The rock is like the fine-grained granite of Mount Desert Island and probably constitutes part of a single body which extends beneath Frenchman Bay. On weathered surfaces the granite displays abundant small, irregular cavities called weathering pits. Many of these may have been initiated by original tiny cavities in the granite.

Here the granite has been broken along huge, slightly curved fractures called sheeting fractures, which incline gently toward the sea. In addition it is broken by numerous steep fractures or joints. In combination, the steep joints and the pronounced sheeting fractures have caused the ledge to break off in gigantic slabs, leaving extensive, nearly flat areas with occasional steps. Where steep joints are numerous, a more rugged and blocky surface is created. Many of even the larger blocks have been moved about and in part further broken up by the surf. A pronounced shift in the location of the large blocks along the shore here may be noted from one season to the next.

An outstanding feature of Schoodic Point is the abundance of dark-colored basaltic dikes, which range up to many yards in width and run through the granite ledges in a roughly north-south direction. Many are severely fractured or jointed, more so than the adjacent enclosing granite. In such instances the small blocks of dike rock may become flushed out and removed by wave action tending to create deep chasms in the more massive granite. A number of examples may be seen where basaltic dikes have been removed by this action for many feet below the surrounding granite, and in some the rubble of dike material may still be seen at the bottom of these crevasses.

It may be noted that, where joints are closely spaced in the granite, the extensive removal of granite blocks by wave action and weathering has also led to the formation of a deep chasm, like "Thunder Hole" on Mount Desert Island. In such cases, however, no evidence of a dike is found at the bottom of the depression. Nevertheless, the chasm produced may closely resemble that formed along a dike. One essential difference, however, is generally to be noted. Erosion along a dike is more likely to lead to rather straight-walled chasms, whereas weathering in severely jointed granite may follow a highly irregular course. A splendid example of the latter type is found at "The Raven's Nest," along the shore two miles northwest of the parking lot, where a ragged chasm about 60 feet deep is cut back into the cliff.

Opposite the middle of the lower parking level, two basaltic dikes about six feet and four feet wide extend up from the water through the continuous ledge of granite to the line of stones at the edge of the parking lot. These two dark-colored dikes are not parallel but intersect at a point about 60 feet from the line of large stones. At the intersection note that the smaller dike cuts through the larger; and where they reach the line of large stones, the dikes are again 40 feet apart with the wider dike on the right. Near the intersection the contact of the younger dike cutting through the older is quite irregular. Upon close inspection, however, this contact is indicated by a marked zone of finer-grained rock in the younger dike where it has chilled against the older, colder dike (see section on "Dikes"). Here we must conclude that, not only is the wider dike older than the narrower one,

Sequence of fractures formation in granite. Here a large block of barnacle-encrusted granite was first detached from the ledge largely by wave action. Subsequently the block was divided by joints (fractures) that opened in two roughly perpendicular directions. In time each block will be further subdivided as new fissures develop and extend themselves. Dissection will continue to render the granite more and more vulnerable to wave attack and eventual destruction. Raven's nest area.

but the time lapse between the two intrusions was sufficient for the older dike to become relatively cold.

At the eastern edge of the parking lot is a trail leading to the rest rooms. Walk 60 feet along this trail, and then turn right and walk to the bare granite ledge on the shore. Notice the small dark-colored basaltic dike running directly down to the water's edge. Near the trail this dike is only a few inches wide. It broadens to two feet toward the water. Near the thin end, the dike is disrupted and offset in distinct segments. Examine this part of the dike carefully and note its banded character. It appears to be composed of thin layers, which curve around roughly parallel to the margin of each segment. In the thicker uninterrupted portions the layers are very regular and parallel the dike walls. The layers are presumably due to flowage of the basaltic melt when it was intruded along the dike fracture. A second similar dike lies a few yards to the east and pinches out toward the water.

Walk westward along the shore until you come to the high metal fence. Just before reaching the fence, you will pass over an unusual dike about 25 feet thick. Unlike the other dikes, the material of this dike is light gray colored on the weathered surface but nearly black where freshly broken. On the weathered surface one may see multitudes of delicate lines or stripes in parallel arrangement. These features, in reality thin layers, formed in response to fluid flow along the dike wall before the dike magma solidified. The slightly turbulent nature of this flowing rock-melt is revealed by the complex wavy pattern and undulations of the flow layers. Look for the small inclusions of granite enclosed in this dike and note how the flow layers wrap around them in streamline fashion. This dike rock is vastly different in composition from that of the basaltic dikes so common along the shore. It carries abundant small crystals of feldspar in an extremely minutely crystallized matrix and is known as a porphyritic felsite. Strikingly enough its composition is very close to that of the granite through which it has cut. If the contact between the felsite and granite can be located, it will be seen to exhibit an extremely irregular outline compared to that between the granite and basaltic dikes.

A deep narrow chasm formed largely by wave action along a basaltic dike near Schoodic Point.

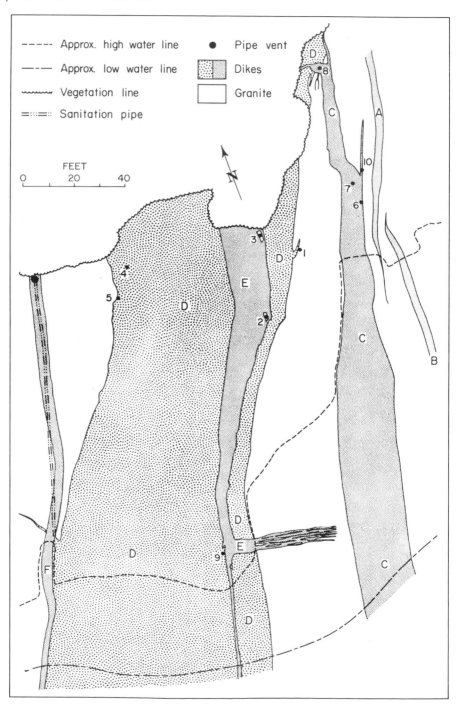

Approx. high water line

Pipe vent

Approx. low water line

Dikes

Vegetation line

Granite

Sanitation pipe

FEET

0   20   40

4. THE HUGE BASALTIC DIKES

Walk back along the shore to the east (to your left as you face the water) until you pass over the sanitary waste pipe. Just beyond are a number of large basaltic dikes. A sketch map of these dikes is shown on the opposite page. Notice that the north direction of the map is toward the top of the figure, the sanitary waste pipe runs along the left side of the map, and the scale is given in terms of feet. A little study of this map will enable you to locate and observe many interesting and striking features about the dikes. For convenience the dikes have been lettered A — F.

Like many of the dikes on this part of the Maine coast, these dikes run a little to the east of north and, as we shall see, appear clearly to fill enlarged fractures in the granite. As was pointed out in the section on "Dikes," most basaltic dikes exhibit extremely fine-grained margins that pass gradually into progressively coarser rock nearer the center. These finer margins are a consequence of rapid or "forced" crystallization due to sudden cooling or chilling of the basaltic melt in contact with colder enclosing rock. Such fine-grained margins are commonly called "chill zones" and serve to indicate that the dike formed from a melt and that it is younger than the rock against which it is chilled.

At (1) on the diagram is a curved offshoot from dike D that penetrates the granite for several feet before pinching out. At (5), on the opposite dike wall, tiny stringers of the same dike penetrate the granite. A very straight offshoot leaves dike C at (10) and cuts the granite. At its junction with the main dike, this offshoot is four inches wide. It gradually thins down and terminates about 20 feet beyond. These small offshoots or apophyses (outgrowths) project well into the granite and serve to indicate that the dike is younger than the enclosing granite. Such small apophyses will be found to have chilled clear to their centers, further substantiating the idea that the dike rock is younger than the adjacent granite.

At (6) are several slabs and irregular blocks of granite, within dike C, aligned parallel to the dike wall. The blocks appear to have been torn loose from the granite wall and engulfed by the basaltic melt. These blocks or inclusions, therefore, testify to the fluidity of the dike material as well as to its age relative to that of the granite.

Along the east side of the map dikes A and B appear to overlap one another slightly. One tapers and pinches out about where the other begins to widen. Immediately to the west is dike C which attains a width of about 25 feet where it disappears beneath the water. Northward it thins down and nearly pinches out before it reaches the line of vegetation. Here (point 8) the dike forks, and one branch turns abruptly at right angles and cuts into dike D. Dike D, therefore, is older than dike C.

Dike D, furthermore, is the largest dike in the area and in places attains a width of 80 feet. Actually this dike consists of two portions separated by dike E. Dike D is the earlier and was later cracked open to admit the basaltic material for dike E. With the aid of this diagram you may actually locate and trace the two seams that separate these two dikes. Either marginal zone of dike E exhibits a narrow layer of finer-grained rock against the older dike D. Here again we have evidence that the thinner dike E was emplaced after the large dike D was relatively cold. Not only was the older dike D cold at the time when dike E was intruded, but it was sufficiently brittle to fracture and shatter so that a new supply of hot basaltic melt could intrude it. Evidence for this is found at (2) where a three foot block of older dike D appears slightly dislodged from the east wall of the fracture and engulfed by the hot basaltic liquid which formed dike E. At (3) is a similar block or inclusion about six feet long. Note that the hot melt, coming in contact with each immersed block of dike D, was suddenly chilled to form a very fine-grained zone in the dike against these inclusions.

If we trace dike E toward the sea, we will note, particularly along its eastern contact, numerous unevennesses and re-entrants in the dike wall. Many of these may have formed as blocks of dike D were dislodged and engulfed. At (9) a block of dike D appears to be caught in the act of being pried loose into the younger dike E. Here also dike E has thinned down to about 4 feet. It suddenly forks, and a thick branch turns abruptly to the east and runs out to the edge of the cliff. A much thinner branch continues southerly under the sea. If you climb down over the cliff here and trace the easterly branch, you will find that it subdivides into a multitude of thin interconnecting veinlets, which terminate 30—40 feet beyond. It is considerably more than coincidence that the easterly branch of dike E (near # 9) has a counterpart in dike C which also cuts the older dike D (at # 8). It seems likely that dikes C and D with their divided thinner ends formed contemporaneously.

At (7) is a light gray, fine-grained, altered granitic inclusion about three feet long and eight inches wide. Examine this material and note that it differs from other granitic inclusions. Furthermore the basaltic dike in contact with this inclusion does not appear chilled against it.

Looking north from Little Moose Island to the Anvil on Schoodic Peninsula. In the foreground a dark basaltic dike, several feet wide, cuts obliquely through the light colored ledge of fine-grained granite.

Presumably, therefore, this inclusion is different from the others and probably represents a granitic block incorporated at an early stage, perhaps at greater depth, while the basaltic melt was sufficiently hot to recrystallize it thoroughly.

At (4) thin stringers (veins) will be found cutting through the basaltic dike D. These veins are rich in a light green colored mineral, epidote, and were formed after dike D had solidified.

Return to the north end of dike C (8) and walk northward along the shore (keep the water on your right). You will soon come upon a large basaltic dike gradually emerging from the line of vegetation. This is probably the northern extension of the dike D-E combination seen further back. About 400 feet north of (8) is a large high ledge standing well above the blocky beach material along the shore. Walk up onto the ledge and notice that you are standing on a younger dike (25 feet wide) of slightly different rock. This dike runs obliquely to the shoreline here and cuts clearly through the older dike which parallels the shoreline. The younger dike rock is also basaltic but carries abundant whitish crystals of feldspar (up to an inch or more long) many times larger than any other mineral grains in the rock. Such a rock is said to have a porphyritic texture and is known as a porphyritic diabase. Dikes of such material are not abundant in the Schoodic Point area. The rock is darker colored and somewhat coarser grained than that of the more common dikes.

5. SCHOODIC POINT TO BAR HARBOR

Note the mileage reading on your car as you leave the parking area. Drive back along the main road, past the Naval Radio Station, to the road junction just ahead. On your way you will pass road-cuts in dikes D and E which you saw along the shore. Bear right at the junction onto what is called Wonsqueak Road, until you come to the Blueberry Hill Picnic Area (on your right) one mile from the parking lot. Turn into the picnic area and park.

Granite, traversed by occasional basaltic dikes, may be seen along the shore and on the neighboring islands. Looking southeast toward the ocean, you will see Schoodic Island about half a mile away. To your right and close by is Little Moose Island, which at low tide is connected to the mainland. A brief visit to this island to see the severely jointed (fractured) granite shore and the numerous basaltic dikes is quite worthwhile.

Behind you and across the road to the north a few hundred yards is a small hill with steep slopes known as "The Anvil." Such a conspicuous elevation is due to the very small number of joints in the granite here, which makes the rock quite resistant to erosion. Out from the foot of the steep slope the granite was broken by abundant fractures and was, therefore, much more susceptible to erosion.

Along the road, 0.6 mile north of the picnic area, is a small turnout (6 on the Schoodic Head map). Park here and walk past the line of large stones toward the shore. The steep slope down toward the water here is held up by granite like that on Schoodic Point but much

A peculiar surface formed on basaltic dike rock at Schoodic Point. The bizarre pattern is due to differential weathering in which the rock is irregularly decomposed and disintegrated. Well-defined fractures (joints) in the rock appear to have been widened and sharp corners appear to have been rounded in the process.

finer grained. Another noticeable difference is that this granite is chock-full of inclusions of slightly darker colored, fine-grained rock. The inclusions range up to many feet across and are irregular to round in shape. They appear to be fragments of older volcanic rocks that were disrupted and engulfed by the fine-grained granite. Unfortunately the undisturbed volcanic material into which the fine-grained granite intruded is nowhere to be seen on the whole Schoodic Peninsula. Only small masses of these volcanic materials, in the form of inclusions in the granite, have been preserved along this part of the coast. There is good reason to believe, however, that at one time the granite of Schoodic Peninsula was completely roofed by the volcanic rocks. The included fragments in the fine granite here have been so reacted upon and altered by the hot granitic melt that they bear little resemblance to the original volcanic material from which they came. Through partial solution and recrystallization, these foreign materials have come to look more and more like the enclosing granite. The very fine texture of the granite here, as compared with that of the granite at Schoodic Point, may in large part be due to the pronounced chilling effect of so many cold inclusions in the granitic melt. Look to the right along the shore and see the closely spaced sheeting joints in the granite ledge dividing the rock mass into conspicuous sheets or layers, which incline at a low angle toward the ocean.

Stop along the road about two miles from the picnic area (at 7) and examine the ledges along the shore. These appear to be composed of fine-grained granite crowded with rounded to long thin masses of altered basaltic rock. Locally so numerous are the basaltic inclusions that there is room for only a small amount of granitic material.

Just ahead is Wonsqueak Harbor, a deep narrow harbor developed along a severely jointed zone in the fine-grained pink granite. Continue on about two miles to the junction with Route 186. Bear right at this junction to Prospect Harbor (about two miles) and turn left on Route 195 to Route 1 near West Gouldsboro. Retrace your route to Bar Harbor.

Erosion of a basaltic dike resulting in a straight-walled chasm on Schoodic Point.

PART IV

BY BOAT TO ISLE AU HAUT

Western Head Shore, Isle au Haut.

1. SHOULD YOU VISIT ISLE AU HAUT?

Before going to Isle au Haut, look at the map to prepare yourself for what you might be able to see there. The island is six miles long (north-south) and 2.5 miles wide. Topographically it consists of a central mountain ridge flanked by a somewhat steeper eastern than western slope. At the south end are two peninsulas, Eastern Head and Western Head, which trend roughly north-south. On the west, near Moore Harbor, is a large peninsula which forks at its south end. North of Eastern Head is a deep north-south furrow occupied by Long Pond. This strong north-south pattern in the topography of the island is merely the reflection of the underlying bedrock architecture.

From the map we see that Isle au Haut is composed principally of four rock units arranged more or less in north-south belts. On the east is a narrow belt of dark crystalline rock that we will call diorite. In the center, holding up the high ground, is a wide belt of fine-grained granite. Between is a very narrow zone of intermediate rock. On the west, volcanic rocks make up the fourth and somewhat discontinuous belt. On Kimball Island, immediately to the northwest, granite is most abundant; but a small strip of diorite follows the shore along the southeast side. The diorite and fine-grained granite correspond to the same rock units on Mount Desert Island, and the volcanic rocks are believed equivalent to types in the Cranberry Island Series.

The island is nearly completely forested, and approximately half of it lies within Acadia National Park. There are numerous trails, particularly across the central portion, but no attempt should be made to walk these without first checking their negotiability with Park authorities or local residents. Certain trails shown on maps of the area are now more or less overgrown or otherwise unsuitable. One auto road, much of it improved, leads completely around the island. A few branch roads make other parts of the area accessible.

The potential visitor to Isle au Haut is warned that he is coming to a relatively isolated and primitive part of the country. He will find Isle au Haut village a small settlement of scattered houses in the northwestern portion, and a number of small farms and widely separated homes along the main road. The village has one small general store where staples may be obtained only between certain hours and on certain days, but there is neither restaurant nor hotel. From time to time one of the natives may offer room and board through the summer months, but accommodations should be arranged before coming to the island. At present the National Park has no provisions for overnight camping; however, one might find some native who will permit such activity on his private property. A stay of two nights should be adequate for one to see the highlights of the scenery, the natural life, and the geology.

The inaccessibility and remoteness of Isle au Haut and the lack of everyday comforts and conveniences of living may not appeal to many; for those a visit to the island should not be undertaken. But for those others who enjoy a taste of the primitive way of life in a setting of scenic beauty and relaxing tranquility, a visit to Isle au Haut will be found highly rewarding.

The excursions on Isle au Haut may be undertaken on foot, by bicycle, or by auto. They all start from the town landing. One way proceeds counterclockwise around to the south side of the island. The other takes you clockwise, on a better road, around the north end to the southeast extremity. For the hiker, each trip will require a full day. If you bring a bicycle to the island, the trips may be made more leisurely and there will be ample opportunity to pursue additional activities which may suit one's fancy. Auto rides accepted from the friendly islanders will greatly expedite your journey between remote points. In any event, it would be quite unwise to attempt to assimilate in a single day all that is described here.

2. ON THE BOAT TO ISLE AU HAUT

Isle au Haut can best be reached by boat from the village of Stonington. Follow state Routes 172 and 15 from Ellsworth, which will take you by bridge to Little Deer Isle and by causeway to Deer Isle at the south end of which lies the old quarry town of Stonington. Transportation to the island is provided by the Isle au Haut mailboat, and the trip requires some 45 minutes. Boats leave from the mailboat landing in Stonington (except Sunday) at 7:00 and 10:00 a.m. and return at 8:00 and 11:30 a.m., but this schedule should be checked for possible changes. Special trips by request will be made at a substantially higher cost. There are restaurant and room accommodations in Stonington, but those who wish to stay overnight here and leave on the early boat must make special provisions for breakfast because at that hour no restaurant is likely to be open.

After leaving the mailboat landing in Stonington, one passes numerous islands composed of granite much like that on Deer Isle. On a number of these islands the granite was worked during the early part of the century, and small abandoned quarries may still be seen there. Looking back at the village you may see some of the old workings high on the hillside. Ahead on the right the boat will soon pass Crotch Island with some of the largest and most active quarries in recent decades. Quarrying operations have been relatively active on this island since about 1870; from here has come stone for many large buildings throughout the northeastern states. The granite used in building the new Park Headquarters at Hulls Cove came from north of here.

As you pass close to the islands note how the surfaces of the pink granite ledges slope smoothly down to the water. Some of the smaller islands, particularly those nearly devoid of vegetation and soil, have a highly symmetrical domelike form. These former hilltops, having been swept nearly clean of cover by rainwash and waves, reveal the smooth rounded forms produced by the spalling (cracking) off of thick sheets or slabs parallel to the surface. This phenomenon of grand-scale spalling, in sheets parallel to the land surface, is called sheeting and is explained in the trip from "Paradise Hill Overlook to Cadillac Mountain" on Mount Desert Island. From time to time you may see sheets and sheeting fractures, all inclining toward the water.

Note that except locally the ledges of the islands are relatively free of joints, but where they have developed, the joints appear to have formed largely in a single direction. The manner in which the granite weathers along joints is a characteristic feature. Disintegration of the rock takes place so as to produce rounded edges and corners and wide open fissures or channels along the fractures. Sharp angular corners on ledges or isolated blocks are rare.

This combination of weathering and sheeting in rock so free of joints accounts for the smooth rounded ledges in the Stonington region. The scarcity of joints also explains why this granite makes excellent quarrying material. Compare the ledges of this rock with the highly fractured, rough, and blocky appearance of the granite that you will soon see, near the north end of Isle au Haut and on Kimball Island to the west, just before you land. Partly responsible for the rounded form of the granite surfaces is the erosive action of glacial ice. Note the great number of large granite boulders in the island area, some perched high on the bare ledges. Originally angular blocks, these erratics were rounded partly by abrasion during glacial transport and partly by later weathering and disintegration.

A few miles out of Stonington you may be able to identify some of

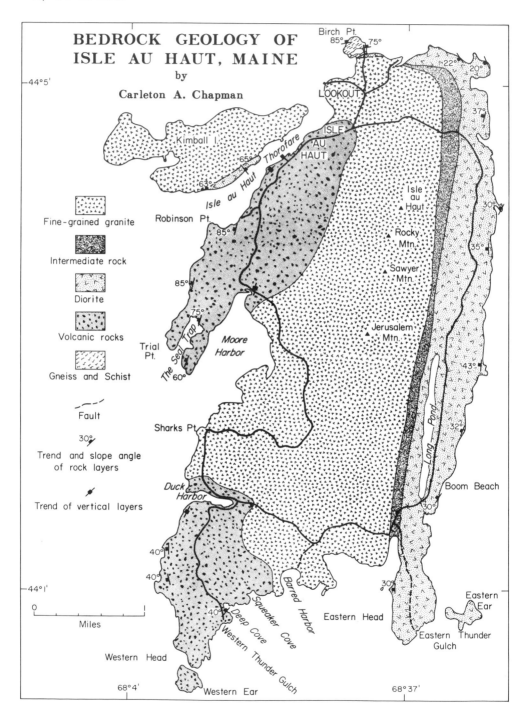

the larger and distant islands. Ahead to the southeast is Isle au Haut, which, compared to the flat islands of the region, stands like a mountain in the ocean. To the east, about 8 miles away, is Swans Island; and at about the same distance to the west are the islands of Vinalhaven and North Haven. Under favorable weather conditions you should be able to spot the high peaks of Mount Desert Island about 25 miles away to the northeast.

After passing through relatively open water for a mile or so, the boat enters a group of small islands and comes close to the north end of Isle au Haut. It may make a brief stop at the small summer settlement of Lookout before entering Isle au Haut Thorofare and tying up at the town landing. Along the eastern end of Kimball Island, near the entrance to the Thorofare, are a number of partially excavated Indian shell heaps. The white shells stand out clearly against the grassy pasture land.

3. WEST SIDE OF THE ISLAND

Robinson Point and the Lighthouse.

Walk up from the Town Landing and take the main road to your right. About 0.4 mile ahead take the road to the right. Roughly 300 yards down this road you will come close to the shore. By following the shore around to your left as you face the water, you will come to Robinson Point and the lighthouse.

As you walk along the shore to the point, notice the material making up the rock ledges. It is a medium to light gray volcanic tuff in which you may distinguish what were originally fragments up to three inches across. Observe in the ledges ahead that these larger fragments in the tuff appear stretched thin in a direction about 10° east of north. Mention of this direction will be made again later. The tuff is typical of much of the volcanic rocks that formed a roof over the thick sheet of granite on Isle au Haut.

From the high ledges around the lighthouse it is but half a mile northward across Isle au Haut Thorofare to Kimball Island. Looking westward from the lighthouse on a clear day, one may see Vinalhaven Island six miles away.

Overleaf: Looking south-south-east to Eastern Head and Eastern Ear (island). Long Pond fills a depression formed within the diorite close to its contact with the fine-grained granite.

Some 15-20 yards north of the lighthouse the gray ledges of volcanic rocks carry abundant round masses of darker colored material several inches across. Locally these masses have a greenish color where they have been converted in part to the green mineral epidote. Although the shape of the masses resembles a type of volcanic fragment known as a volcanic bomb, it is doubtful that these bodies are such features. They appear to represent portions of the tuff in which certain darker colored minerals concentrated sometime after the tuff was deposited.

Volcanic tuff is encountered in the ledges along the shore to the southwest all the way to Trial Point (1.3 miles from the lighthouse). Locally the conspicuous larger fragments in the volcanic rock along here reach the length of many inches and the rock can justifiably be called volcanic breccia. At the head of The Seal Trap much good breccia is exposed bearing fragments up to a foot across. This breccia forms distinct layers or beds alternating with beds of the normal tuff. Here the beds have been tilted so they incline steeply to the west. A careful measurement shows that these beds, on a horizontal surface, trend about 10° east of north. This direction, as indicated by both bedding and elongate fragments of the volcanic rocks, takes on definite meaning if we now examine the map. This direction is the same as the trend of the peninsula between Robinson Point and Moore Harbor. The directional structure of these volcanic rocks was sufficiently pronounced to influence the shape (long and slender) and direction of the whole peninsula. The Seal Trap is but a minor expression of the control exerted by this rock structure. Here a zone in the volcanic rocks, weaker than that on either side, has permitted deeper erosion of the earth's surface and the creation of a small bay flanked by two peninsulas, all trending 10° east of north.

Moore Harbor.

If you walk around to The Seal Trap, cut across the small peninsula beyond and follow the shore northward to the road at the head of Moore Harbor. If you did not go far beyond Robinson Point, return to the main road, turn right, and continue to the head of Moore Harbor. About 100 yards from the road is a natural bar built eastward almost across the mouth of the small stream that enters the head of the harbor. The bar is about 10 feet high and 200 feet long. It is composed mostly of cobbles with some pebbles and a few boulders. The higher part is topped with grasses and other low vegetation. The bar has formed here in response to strong currents and wave action created largely by southwest winds.

Walk to the east side of the harbor, just off the main road, and cross the small stream bed there. Just ahead (south of the east end of the bar) are several flat ledges of light gray volcanic rocks. Compositionally these rocks are mostly rhyolite, and on the exposed and weathered surfaces they appear almost white. Freshly broken surfaces, however, appear dark gray due to the extremely fine-grained nature of the material. The rough, jagged surfaces of these ledges are due to the severity of jointing or fracturing of the rocks. Joints are only about an inch apart. That these rocks were once flowing lava is indicated by their extremely fine-grained character and by the delicate striped appearance seen on certain smooth surfaces. Individual stripes are a millimeter or so wide, and adjacent stripes alternate in color between lighter and darker shades. These stripes represent layers that developed in slightly dissimilar hot lava as it flowed in a rather uniform manner. Locally the layers are highly contorted and complexly rippled due to irregularities in the currents of flowing lava. If you examine these ledges carefully, you may see that locally the rhyolite is rich in tiny spherical bodies up to a few millimeters across. These are the so-called spherulites (star-shaped clusters of needlelike crystals) so common in rhyolite lavas from all over the world. On some flow-striped surfaces you may discover abundant tiny angular fragments embedded well within the rock. In most instances the long dimension of the fragments is oriented parallel to the flow layers indicating that the fragments were incorporated by the molten lava when it was sufficiently fluid to develop the flow layers and to turn the fragments into streamline positions in the flow. Some fragments may have been caught up in the lava as it spread laterally over the ground; others may have entered the lava from above as fallen material discharged from neighboring erupting volcanoes.

Traversing the ledges here are a few irregular dikes of felsite up to one or two inches thick. They are compositionally like the rhyolite and probably represent fractures in the more brittle outer parts of the lava flows that opened slightly and filled with hot melt from the still liquid interior.

About 150 yards south of the east end of the bar (under the trees, at the vegetation line along the shore) are smooth ledges of volcanic rocks exhibiting highly polished surfaces. Looking at these surfaces in the proper light, one may see they are traversed by numerous striations (fine grooves) and deep scratches, which trend in a north-south direction. Such polished and finely grooved surfaces are quite common over the island, even well above the water line. They represent abrasive action by glacial ice impregnated with abundant rock particles. From this observation we can conclude that, at the time these scratches and striations were produced, the glacier in the vicinity of Moore Harbor was moving due south.

Instead of returning to the main road, one should follow the shoreline south, along the east side of the harbor, for several hundred yards. This will bring one into the Park area close to the main road and will provide a splendid opportunity to study the nature of the contact between the volcanic rocks and the granite of Isle au Haut.

For about 250 yards southward from the polished and striated ledges, one will pass over more of the volcanic rocks. Next one crosses a zone (about 350 yards wide and extending to about the Park boundary) composed of mixed rocks (volcanic and granitic). At first a few scattered dikes of very fine-grained granite are seen cutting through the ledges of volcanic rocks. These become more numerous to the south, and the volcanic rocks appear to become more thoroughly broken into smaller and smaller angular blocks, which are recemented in a matrix of very fine-grained granite. Further to the south granite makes up most of the ledge, and the size and abundance of enclosed rock fragments decreases. Near the point where the Park boundary comes down to the shore, inclusions of volcanic rock are still visible in the granite.

This mixed zone, across which you have traversed, must have developed when the body of granite on the east was emplaced (set) against the volcanic rocks on the west. Since the granite encloses fragments of the volcanic rocks and since granite dikes cut into the volcanic rocks, we conclude that the granite is the younger material. Space for this granite must have been provided in part by disruption of the older volcanic rocks; and as the granitic melt was emplaced against the shattered volcanic material, it entered fractures in the latter to form the numerous dikes and incorporated abundant volcanic rock fragments to form the wide belt of inclusions in the granite.

It will be noted that in general the blocks or inclusions of volcanic rock enclosed by granite become more rounded and tend to blend in more with the granite toward the south. Such relations indicate that these inclusions well within the granite body have been more thoroughly heated, recrystallized, and chemically attacked by the hot melt and have, therefore, come to resemble more closely the enclosing granitic rock in mineralogical composition and texture.

But just as the granite melt left its effect upon the volcanic rock inclusions, so the inclusions modified the adjacent granitic material. The decrease in grain size of the granitic rock from south to north across the mixed-rock zone is ascribed to the chilling effect of the progressive-ly more abundant cold rock fragments upon the granitic melt. As we have seen, on a much smaller scale, in the case of basaltic dikes, when molten rock material comes suddenly into contact with cold rock, the former loses heat rapidly and is "forced" to crystallize quickly, pro-ducing a much finer texture.

Moore Harbor to Sharks Point.

Continue along the shore to the small cove. Then turn left and walk through the woods about 80 yards to the main road. By following the road to the right for 0.6 mile you will climb a long slope to where the road turns slightly to the right and crosses a small stream or swampy run. From here the road continues to rise, but slightly, and then nearly levels off. You will come next to a swampy area on your left, which gradually approaches and crosses over the main road. Here on the right of the road at the swamp's edge is a ledge of the granite. The rock is buff to gray-colored, fine-grained, and somewhat weathered. Close in-spection will reveal mostly buff to pink feldspar, much glassy-looking quartz, and a little dark-colored hornblende (slightly altered). An in-spection of the map will show that here we are at a considerable dis-tance from the volcanic rocks; the granite, therefore, is coarser than that we found on the shore.

Ahead the road begins to climb again gently; and about 0.3 mile beyond the swamp, it reaches its highest point. Just beyond the crest, on the right, is a small gravel pit, which served as the source for much of the road material used here. Although now partly overgrown, the nature of this deposit may still be seen. The material is composed of sand and gravel arranged in flat layers or beds. Some layers consist of thinner beds, which are inclined at a pronounced angle and are known as cross-beds. Gravel layers are composed largely of pebbles and cobbles, but a few boulders up to two feet across are present. The bedded nature of this material indicates that it was deposited in water, but the subangular or partly rounded shape of the pebbles and cobbles suggests that the material has not been transported far by streams. Pre-sumably these deposits formed when material brought in by the glacier was picked up and redeposited by water from the melting ice. The cross-beds observed here are evidence of water-laid material. Such beds are characteristic of delta deposits and are formed where sediment-laden streams enter bodies of relatively quiet water.

About 1.1 miles ahead, the road comes nearly to the shore. Leave the road and walk out on the beach. To your left, as you face the water, is Sharks Point. Note that the rock in the ledges along the shore is very fine grained, much finer than that further back along the road. Checking our position on the map, it appears that we are now very close again to the contact with the volcanic rocks. The contact may lie only a short distance off shore. This accounts for the finer texture of the granite here.

At the head of the cove is a beach of flat pebbles and cobbles composed mostly of very fine-grained granite and only to a small extent of dark gray felsite. The preponderance of granite pebbles is in keeping with the idea that the beach material is derived largely from the local ledges.

The granite pebbles appear darker colored than does the granite of the large ledges. If a granite pebble is broken open, it will be found to be composed of very fresh rock. In such material the distinction between individual mineral grains is difficult to make because the fresh quartz and feldspar look so much alike. The lighter color of the granite in the ledges here is due to partial alteration of the rock, and where weathering has been most intense, the feldspar takes on an opaque

Western Head, Isle au Haut.

white or buff color. This gives a much lighter cast to the whole rock. Weathering tends to weaken rock material, thus we can understand why these pebbles, which suffer constant wear on the beach, tend to be composed of the freshest and therefore most durable material.

The felsite pebbles, though dark gray on their smooth surfaces, may be nearly black on freshly broken surfaces. This material represents some of the freshest to have come from the volcanic rocks to the north. Compositionally it is rhyolite, like that at the north end of Moore Harbor. On some pebble surfaces the flow layers and spherulites described above may be detected. Furthermore, the smooth pebble exteriors may reveal the presence of numerous tiny white crystals of feldspar and quartz.

On smooth pebble and cobble surfaces one may see numerous tiny bowed or crescent-shaped scars. These little cracks or percussion marks were created by intense local impact with adjacent pebbles and are continually being formed as the beach materials are rolled and tossed about by the waves. The formation of myriads of tiny cracks in the surface layer of such fine-grained material tends to diffuse the light falling upon the pebble surface and to give it a somewhat lighter color.

Sharks Point to Deep Cove and Western Thunder Gulch.

Return to the main road and continue on to Duck Harbor, about a mile away. At the head of the harbor the road branches. Take the right branch, which leads down across the bridge and along the south side of the harbor. Duck Harbor is an interesting physiographic feature. It occupies a deep narrow valley that trends east-west and that was formed presumably by erosion along a narrow zone of severely fractured rock.

Continue on the road along the south side of the harbor until it turns suddenly to the left and climbs up out of the valley. It crosses rather level ground for slightly over a mile till it comes down to the coast again at the head of Deep Cove. Do not confuse this cove with another of the same name, north of Shark Point, at the mouth of Moore Harbor. About 500 feet beyond the road ends; and here a path leads straight ahead for 100 feet to the head of a deep chasm, extending back from the water's edge. This is Western Thunder Gulch. It is about 50 feet deep, 200 feet long, and 10-25 feet wide and has developed along a narrow zone of intense fracturing in the high bluff of gray volcanic rock. The closely spaced fractures have so weakened the cliff here that erosion has progressed rapidly. Through frost action, the wash of running water, and the action of ocean waves, the rock material in this shattered zone was readily dislodged, further disintegrated, and swept away leaving a deep narrow depression between high steep walls. In essence this gulch resembles Thunder Hole on Mount Desert Island.

Work your way down into the head of the chasm being careful not to slip on the wet rock surfaces. Just below the line of vegetation in the floor of the upper part of the chasm, you will see a basaltic dike ranging up to ten feet wide and trending about 20° west of north, parallel to the gulch itself. This dike is badly fractured, and by disintegration has aided in the formation of the gulch. When we consider the intensity of fracturing here in the volcanic rocks, however, it seems likely that a gulch would have formed had there been no dike at all.

The gulch derives its name from the pronounced rumble created by the insurging waves. Even during periods of moderate surf, the salt spray rises to the top of the gulch. It should be noted that this gulch, which is on the peninsula known as Western Head, is not to be confused with a similar feature of the same name on Eastern Head only two miles to the east.

Return to the road at Deep Cove and follow it back to the shore. Walk out onto the beach and turn right. Ahead to the south is a broad, gently sloping ledge of the volcanic rocks. The rock is light to medium gray and is composed largely of felsite, but much is volcanic tuff. Some of the felsite contains numerous tiny but conspicuous crystals of quartz and feldspar. Here the layers, or beds, can be seen to trend northeast-southwest and to incline about 40° to the northwest.

A basaltic dike, about 2.5 feet wide, emerges from beneath the cobble beach here and runs southward for about 100 feet where it narrows and finally pinches out. At several places this dike and the enclosing volcanic rock have been interrupted and offset along fractures in the ledge. Along the dike walls, typical thin chill zones have formed; but along the lines of offset where the dike is in contact with granite, no chill zones are to be found. This is evidence that the offset occurred antecedent to the formation of the chill zone and perhaps after the entire dike had crystallized.

Proceed 100-200 feet further south, past the second basaltic dike, and note the volcanic tuff containing tiny angular fragments. These fragments, some of which are three inches across, are best observed on the whitish exposures above high tide level.

Now, reverse your direction and walk along the shore around the end of the cove. You will soon come to a high ledge of light colored volcanic rock known as volcanic breccia. It is composed of angular to sub-rounded rock fragments up to many inches across embedded in a blue-gray to white matrix of extremely fine-grained material. The fragments are varied in color, and many are finely striped like the felsite at Moore Harbor. This striped material was derived from rhyolitic lava flows that were thoroughly disrupted by volcanic explosions. Some blocks in this breccia attain the length of three feet. Further along the shore is more breccia interlayered with the finer volcanic tuff.

Barred Harbor.

If time permits continue along the shore. This will take you over beautiful whitish exposures of the volcanic rocks, past Squeaker Cove, to a wide zone of mixed rock much like that at Moore Harbor. Many of the smaller inclusions in this zone appear to have been intensely soaked by the granitic melt. So thoroughly have they been changed that many are difficult to distinguish because they blend in so well with the enclosing granite. On the point west of Barred Harbor, numerous gigantic inclusions of the volcanic rocks may be observed within the granite.

It is particularly interesting to note that the granite on and around the point is relatively coarse-grained for material so close to the contact with the volcanic rocks. Two additional features makes this granite unusual. It is rich in tiny cavities apparently lined with minute pink feldspar and gray to white quartz crystals, and it contains conspicuous patches of very coarse-grained material called pegmatite. Individual patches of the pegmatits are most generally composed of pink feldspar, and the centers are made up of milky gray quartz.

Tiny cavities, of the type seen here, are generally believed to have formed in granitic melts rich in dissolved water. As the rock melt crystallizes, tiny interspaces in the crystal aggregate become filled with a water-rich material. Mineral grains along the walls of these interspaces continue to grow without interference against the adjacent fluid and thus develop well-defined crystal faces. The associated patches of pegmatite here are also considered evidence for a water-rich granitic melt. The presence of the water phase is believed to enhance the growth rate of crystals and thus explain the coarse texture of pegmatite. Perhaps, then, we can attribute the relatively coarse texture of the granite here to a locally high water content of the melt.

Continue on around the north side of Barred Harbor, and you will pass two ledges of the volcanic rocks that represent huge inclusions in the granite. The first lies near high tide line, about 700 feet west of the small stream or run that enters at the northeast part of the harbor. This ledge is traversed by several small veins of very fine-grained pink granite, up to 2 inches wide, which are probably offshoots from the main granite body. Of particular interest are the abundant and well-developed spherulites in volcanic rock (felsite). These round to oval-shaped bodies are up to an inch across and locally so closely packed they make up most of the rock. The spherulites are star-shaped clusters of needlelike crystals, like those at Moore Harbor; but these are much larger and commonly show a distinct concentric structure like that observed in a sliced onion. Some of the large spherulites have been split apart by the small veins of granite.

About 200 feet to the east is a second large ledge of banded felsite clearly exhibiting good flow-layering. The layers show up best on weathered rock surfaces as stripes up to one-quarter inch wide. This ledge is traversed by tiny pink veins of granite and by numerous zones, 1 to 2 inches wide, of thoroughly shattered volcanic rock. These shattered zones in the volcanic rock are evidence of some pronounced disturbance and were probably produced just prior to the emplacement of the granite of Isle au Haut.

Just east of the small stream or run is a foot path that will take you back to the main road about 500 yards to the north-northeast.

4. BIRCH POINT

From the Town Landing, take the main road to your left for about 0.7 mile. You will pass through the small village of Isle au Haut and come to a side road leading off to the left. About 300 yards along this side road is a broad turn where 2 branches lead off to the right. Take the second branch, which will bring you out to the shore at the north end of the island. The road passes over a bridge to a small island on which Birch Point is located. Just beyond the bridge turn right to the beach. Then turn left and walk around the island in counterclockwise fashion.

The rocks along the shore here are quite different from any on Isle au Haut. They show distinct layers composed of different material and trending roughly in an east-west direction. Much of the material is light to medium gray and is composed of quartz, feldspar, and biotite. The rock shows a distinct streaking or foliation due to parallel arrangement of its component minerals. Some rock layers split readily into thin slabs and are called schist. Others, in which this quality is lacking, may be called gneiss. More or less interleaved with schist and gneiss are layers of massive dark gray rock rich in biotite mica. This rock carries numerous pods of green epidote an inch long and is traversed by small veins of epidote and of quartz and feldspar. Locally these darker rocks are badly broken up and invaded by veins of light pink, fine-grained granite. In some places the granite appears as a matrix for angular blocks of the dark rock. Elsewhere it appears to have insinuated itself as thin lenses and sheets along the foliation surfaces of schist and gneiss. Much gneiss and schist carry abundant tiny clots and knots of dark biotite and epidote in ellipsoidal or lenslike forms, elongate parallel to the foliation.

Near the extreme northern part of Birch Point, the gneiss with tiny dark clots is extensively cut by stringers (narrow veins) of pegmatite. This pegmatite is a very coarse-grained, light-colored rock with buff to pink feldspar crystals, up to a foot across, and much milky white quartz. Considerable biotite mica is present as large black flakes, but little muscovite mica, as fine-grained aggregates, will be found. Black tourmaline intergrown with quartz forms masses up to many inches across. On broken surfaces, many large feldspar crystals also can be seen to be intergrown with quartz. Further around Birch Point large portions of the ledges are composed of a basaltic rock resembling the dark-colored dikes found over the island.

Soon you will come to a huge boulder (about 18 feet long, 15 feet wide, and 12 feet high) composed of the Stonington type of granite. It is cut across by a steeply inclined vein (8 inches wide) of light pink, fine-grained granitic rock with a very coarse-grained central portion. The boulder is split and slightly separated roughly along the vein. The granite is made up of glassy gray quartz, pink potassium feldspar, white plagioclase feldspar, and black flaky biotite. The feldspar crystals attain a length of two inches, and many show pink centers of potassium feldspar and white rims of plagioclase.

This boulder of granite, quite different from any in the ledges of Isle au Haut, must have been carried here by the glacier from some point to the north-northwest. Boulders of this type of granite are common all over Isle au Haut and Kimball Island to the west, and the rock material can be quickly spotted by its coarse texture and abundant large feldspar crystals. Boulders of this type of granite, furthermore, are the largest of any found in the vicinity, a perplexing fact when we realize their relatively remote source. One might suspect that the further a boulder travels, the more wear it must undergo, and the smaller it would become. The size, however, depends on several factors; an important one being the size of the initial block from which a boulder is to be formed. We have already seen from the boat that ledges of this type of granite are broken by a relatively small number of joints. The Isle au Haut rocks, by comparison, are cut by abundant fractures. Blocks pried loose by the glacier from the Stonington type of granite should have been many times larger than those from Isle au Haut. This explains the contrasting sizes of the corresponding boulders.

The rock in the ledges near the large boulder is a breccia composed of angular to round blocks of basaltic rock, like that seen farther back, enclosed in a matrix of granite. Much of this breccia resembles that described along the way from Schoodic Point to Wonsqueak Harbor on the return trip from "Schoodic Point to Bar Harbor." About 150 feet south of the large boulder is a basaltic dike about 8 feet wide. It cuts through the gneiss and breccia so it must be younger than any other rock here. A second basaltic dike 6 feet wide cuts the layered rocks 300 feet south of the boulder.

Continue on around the next point of land to where a beach runs nearly east-west. This beach you will notice has water on both sides and is known as a bar. A short distance ahead, the bar is terminated by a very small island; but beyond, another bar begins and extends to Isle au Haut. Thus, it appears that the island on which Birch Point is located, the very small island nearby, and Isle au Haut itself are all in line and tied together by two bars. Bars tying island to island or island to mainland are called tombolos.

On the very small island you may be able to locate the contact between the granite of Isle au Haut and the basaltic rock. This contact runs about east-west and the granite is on the south side. The basaltic rock is cut by small granite veins, and the granite near the contact includes numerous blocks of the dark basaltic rock. During high water much of this ledge may be covered.

Continue along the bar and return to the road. Go back out to the main road and turn left. About 0.6 mile along the main road will bring you to the height of land. On the right just ahead is a large cut in granite. This material is very typical of much of the granite on Isle au Haut.

5. EAST SIDE OF THE ISLAND

Looking at the map we see that on the east side of the island the main road runs entirely along the belt of diorite and at a remarkably constant elevation. In the northern part for a short distance west of the road is flat swampy ground, whereas to the south the road is flanked by Long Pond. Further west the land rises rather abruptly, beginning about at the zone of intermediate rock and culminating in a long ridge or series of small mounts (Isle au Haut, Rocky Mountain, Sawyer Mountain, and Jerusalem Mountain) composed of the granite.

From along the road one may see that Long Pond is located in a narrow and steep-sided valley cut into the belt of diorite. This north-south valley is without doubt following a weak zone in the diorite belt. Very likely the disturbing forces in the earth's crust that tilted all the rocks of the island to the west, as we shall see later, were also responsible for shattering the diorite in the vicinity of Long Pond. Other shattered zones, such as those at Duck Harbor and at the two Thunder Gulches, were perhaps produced at this same time. During Pleistocene time glacial ice must have moved down this narrow valley, which lay nearly in line with the main direction of ice flow. Perhaps some deepening of the valley is to be attributed to ice action, but the general shape was modified only slightly.

Eastern Head and Eastern Thunder Gulch.

About half a mile beyond the south end of Long Pond, the main road comes close to the ocean and then turns sharply to the right. From the turn, walk down to the shore and along the head of the bay to the far (east) side. Follow the shore south for about a mile to the tip of Eastern Head. About 200 yards south from the head of the bay, you will encounter massive ledges of dark-colored diorite showing the effects of glacial wear. Small striations are well preserved on many of the smoothed surfaces, and you may note some rather large grooves cut into the ledges here. Both grooves and striations trend 15° to the east of south, nearly the same direction we noted on the east shore of Moore Harbor. Other interesting features produced by glacial erosion are the so-called roches-moutonnées or rock sheep. These are conspicuous protuberances of the rocky surface with long, smooth, gently inclined north slopes and short, steep, rough south slopes. These bulging forms appear streamlined in the path of moving ice. On the upstream side (north) the advancing ice was able to scour and smooth projecting portions of the ledge. On the downstream side (south), in the lee of the ice current, abrasion was less important; but small blocks of rock were plucked out and removed leaving a steep irregular surface.

Between here and the south end of Eastern Head, you will pass over a variable group of rocks; but for the most part they may be classed as diorite. The fresh material is very dark and relatively coarse-textured and is composed principally of plagioclase feldspar and hornblende. These minerals are more readily distinguished by the layman on a weathered surface because there the plagioclase is light gray or buff whereas the hornblende is almost black. In places, pyroxene or biotite may be important dark constituents. The most massive rock carries large round crystals of hornblende or pyroxene up to an inch across, and these enclose abundant small feldspar crystals. On the weathered surface this rock develops a very characteristic rough, pock-marked appearance.

About 1000 yards south from the bay head is a small point of land jutting out from the rather straight shoreline. Above high tide here may be seen a curved storm beach enclosing a small lagoon on the landward side. The storm beach was formed by material tossed up by high waves, and its shape was controlled by the rocky reef which extends out to the small island just offshore and which is visible at low tide.

Along the shore the ledges appear to be composed of distinct layers of at least two kinds of diorite, a lighter and a darker variety. These layers trend north-south and incline westward under the water. The small island just offshore is composed of granite. Between the diorite and the granite is intermediate rock. It is clear then that the granite and intermediate rock are resting above the diorite because the layers of diorite incline down under the small island.

Throughout the mass of diorite, along the east side of the island, are veins and variously shaped masses of lighter colored rock. Some of this is a type of coarse-textured diorite with more or less quartz, but much has a composition ranging to that of granite.

From the high ledges along the shore of Eastern Head, there is a magnificent view to the west. Whitish cliffs and ledges of volcanic rocks, which start on Western Head, can be traced around nearly to Barred Harbor where the pink granite begins. The view is perhaps best in the morning when the light falling along the shore provides a striking contrast between the dark green of the spruce forest and the deep blue of the ocean, separated by a narrow fringe of white surf breaking against the rocky coast. To the southwest is the Matinicus Island group about 18 miles away.

Continue around the southern tip of the head and follow the shore for about 1000 feet to a small indenture in the coastline. This is Eastern Thunder Gulch. Its dimensions match those of the Western Thunder Gulch on Western Head, but its trend is about 30° west of north. This gulch has developed in coarse diorite along a zone of intense fracturing. Where severely fractured, the diorite becomes highly susceptible to chemical decomposition, which greatly aids the process of weathering. A few yards west of the gulch is a basaltic dike about three feet wide and enclosing fragments of the adjacent diorite up to a few feet long. Notice that this dike is not in a position to affect the localization of Eastern Thunder Gulch. Furthermore, there appears to be no trace of a dike anywhere within the gulch. Climb to the top of the gulch, and take the path north over Eastern Head and back to the main road.

Boom Beach.

Approximately 350 yards north of the south end of Long Pond, take the trail which leads eastward from the main road for 150 yards to the shore. Here you will see a deposit (about 400 feet long and 100 feet wide) composed of well-rounded boulders and cobbles mostly one to two feet across. A few boulders range up to five feet through. This accumulation must be of local derivation because over 75 percent of the boulders are composed of diorite like the ledges nearby. The remainder are of granite like that immediately to the west. Some bouldery material has been piled high here to form a storm beach. This boulder deposit rests on a smooth, seaward-sloping rock surface. The booming sound from which the beach derives its name is created as the boulders are rolled across the rocky pavement under the action of a strong surf.

If you follow the coastlike either north or south from Boom Beach, you will encounter more diorite, which exhibits many of the features observable on Eastern Head.

When you leave the shore, walk back out to the main road and turn left. The paved road before you is relatively level and is the best route back to Isle au Haut village and the town landing roughly five miles ahead.

APPENDIX 1. THE GEOLOGIC TIME SCALE

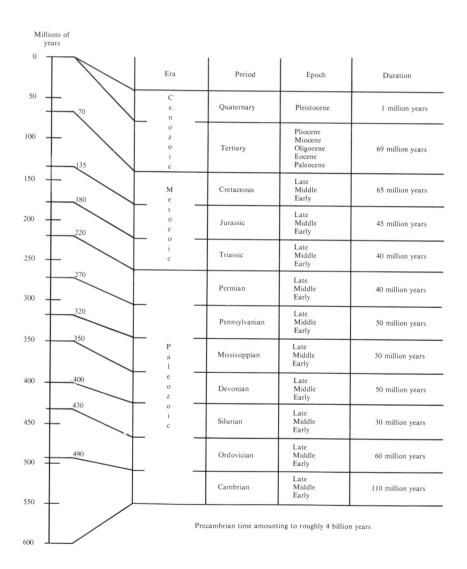

Millions of
years

	Era	Period	Epoch	Duration
	C e n o z o i c	Quaternary	Pleistocene	1 million years
		Tertiary	Pliocene Miocene Oligocene Eocene Paleocene	69 million years
	M e s o z o i c	Cretaceous	Late Middle Early	65 million years
		Jurassic	Late Middle Early	45 million years
		Triassic	Late Middle Early	40 million years
	P a l e o z o i c	Permian	Late Middle Early	40 million years
		Pennsylvanian	Late Middle Early	50 million years
		Mississippian	Late Middle Early	30 million years
		Devonian	Late Middle Early	50 million years
		Silurian	Late Middle Early	30 million years
		Ordovician	Late Middle Early	60 million years
		Cambrian	Late Middle Early	110 million years

Precambrian time amounting to roughly 4 billion years

APPENDIX II

SUPPLEMENTARY READING

The following articles are suggested for those interested in further studies on the geology of Acadia National Park:

Chadwick, G. H., "The Geology of Mount Desert Island, Maine." *American Journal of Science,* vol. 237 (1939), pp. 355-363.

Chadwick, G. H., "The Geology of Mount Desert Island (Acadia National Park)." *New York Academy of Science Transcript,* Series 2, vol. 6 (1944), pp. 171-178.

Chapman, C. A., "Bays-of-Maine Igneous Complex." *Geologic Society of America Bulletin,* vol. 73 (1962), pp. 883-888.

Chapman, C. A. and Rioux, R. L., "Statistical Study of Topography, Sheeting and Joining in Granite, Acadia National Park." *American Journal of Science,* vol. 256 (1958), pp. 111-127.

Raisz, E. J., "The Scenery of Mount Desert Island, Its Origin and Development." *New York Academy of Science Annual,* Number 31 (1929), pp. 121-186.

MAPS

For anyone contemplating an extended tour of Acadia National Park, a large topographic map "Acadia National Park and Vicinity, Hancock County, Maine" is the most useful. This may be obtained at the Park Visitor Center, in local book and stationery stores, or from the U.S. Geological Survey, Washington, D.C. 20242. Topographic quadrangle maps for "Bar Harbor Quadrangle," "Mount Desert Quadrangle," and "Deer Isle Quadrangle" are also available from the same source.

The more casual visitor should obtain a copy of the tourist map of Acadia National Park from the Visitor Center or campsite outlets. Trail maps of Mount Desert Island are also available from the Appalachian Trail Club or the Mount Desert Chamber of Commerce.

Visitors are cordially invited to take advantage of many guided nature trips and campfire programs offered, at no charge, from late June through August by Park Naturalists. The routes which have been selected for these tours include choice examples of flora and fauna as well as many of the geologic features described in this book. Shoes suitable for hiking, preferably with rubber or composition soles, should be worn on all walking trips through Acadia as trails are primitive, ungraded, and sometimes quite steep. Some of these conducted tours and programs are outlined below. As schedules change each year, one should obtain a brochure from the Visitor Center or other Park outlets.

CARRIAGE ROAD NATURE WALK. A four mile walk along carriage roads winding through secluded forest glens. Trips start at the Jordon Pond House parking area.

GREAT HEAD NATURE WALK. Explore the geology and ecology of land and sea with spectacular marine views along a two mile trail starting from the Sand Beach parking area.

SEASHORE INTERPRETATION. Each day during low tide, Park Naturalists are stationed along the shore at Otter Point and Seawall Picnic Area. They are available to answer questions about the marine life and the geologic formations of these areas.

CAMPFIRE PROGRAMS. Illustrated lectures on the natural and human history of Mount Desert Island region are featured at outdoor evening programs held in the Blackwoods and Seawall campground amphitheaters. Other illustrated evening programs are presented in the Visitor Center Auditorium at Hulls Cove.

MUSEUMS. At Sieur de Monts Spring one may visit the Nature Center, the Wild Gardens of Acadia and the Abbe Museum of Indian relics. On Little Cranberry Island is the Ilesford Historical Museum. Ferry schedules may be obtained by phoning Park Headquarters, 288-3338.

BOAT CRUISES. A cruise among the Porcupine Islands and the spectacular cliffs of Ironbound Island makes a delightful morning or afternoon for adults and children alike. For the more venturesome, a trip to Baker Island includes a walk over the island to the lighthouse and the seawall of gigantic granite blocks described in Part II, Chapter 15 of this book. Reservations for these cruises are requested, and details may be obtained from Park Headquarters.

FOR FURTHER INFORMATION on all aspects of Acadia National Park write: Information Office, Acadia National Park, Hulls Cove, Maine 04644.